The Economics of the Modern Construction Sector

Also by Stephen L. Gruneberg and Graham J. Ive

THE ECONOMICS OF THE MODERN CONSTRUCTION FIRM

Also by Stephen L. Gruneberg

CONSTRUCTION ECONOMICS: An Introduction

FEASIBILITY STUDIES IN CONSTRUCTION (*with David H. Weight*)

RESPONDING TO LATHAM: The Views of the Construction Team (*editor*)

The Economics of the Modern Construction Sector

Graham J. Ive
Senior Lecturer
Bartlett School
University College London

and

Stephen L. Gruneberg
Lecturer
Bartlett School
University College London

First published 2000 by
MACMILLAN PRESS LTD
Houndmills, Basingstoke, Hampshire RG21 6XS
and London
Companies and representatives
throughout the world

ISBN 0–333–62667–2 hardcover
ISBN 0–333–62662–1 paperback

A catalogue record for this book is available
from the British Library.

Printed and bound in Great Britain by
Antony Rowe Ltd, Eastbourne

To Rose and Jan

Contents

List of Figures

List of Tables

Preface

The Economics of the Modern Construction Sector is the companion
volume of *The Economics of the Modern Construction Firm,* and in many
ways *The Economics of the Modern Construction Sector* sets the context for
an understanding of the operation of firms within the construction
production process. As economists our particular interest is focused on
the production processes of industrial activity. We are also concerned
with the economic and social environment with which firms interact.

This environment is comprised of the economy as a whole, its construction and other industrial sectors, government and consumers. It is
also determined by the resources available for the production of particular outputs. These are the labour, materials, both natural and manufactured components, and the plant and equipment used. In turn the
productivity of these resources depends on the state of technology and
the methods of production adopted. Finally the behaviour of firms in
the construction sector is also to a large extent determined by the
behaviour of the other firms in the construction sector itself.

We are therefore interested in understanding the economic roles and
relationships within the construction process, in that we wish to
understand how and why production takes place. We also wish to
know who benefits from the system of production and how these
benefits are distributed. The general economic issues raised up to this
point could be applied to any industry or sector of the economy, and
discussion of these and other topics can be found in all introductory
economics texts. However, if the general textbook approach to economics is applied to construction several features of the construction
process intervene to invalidate the conclusions. The result can be a
frustration with economic theory and ultimately a rejection of economics on the grounds that it is neither realistic nor practical.

In this book we attempt to provide a realistic theoretical economic
framework for understanding the construction sector in particular, in a
particular country, the UK, and at a particular phase of its development,
at the end of the twentieth century. What may be true of the construction sector here and today, does not necessarily apply to the construction sector somewhere else and in another period. Armed with this
approach, it should be possible to discuss policies related to construction and account for the behaviour of firms within the construction

sector. Nevertheless, the methods of analysis and the general approach to the economic issues of the construction sector in this book should be a useful starting point for the analysis of any construction sector or construction firm working anywhere in the world. This book therefore provides the necessary background for firms operating in construction and together with the *The Economics of the Modern Construction Firm*, it is hoped that *The Economics of the Modern Construction Sector* will enable firms to improve their decision making, strategic thinking and planning effort.

The four parts of *The Economics of the Modern Construction Sector* are entitled Production, Accounting for Production and Assets, The Nature of the Construction Process, and finally, Construction and the Economy. In the first chapter we begin by defining the production of the built environment and then move on to examine the process of adding value to inputs. The value added approach highlights the source of wages and profits. In Chapter 2 the industrial relationship between construction labour and capital is used to show the subsumption of construction labour by capital. In Chapter 3 the argument is developed to show how labour is used to raise productivity and produce profits.

In Part II, we are concerned with a meaningful statistical analysis of the data surrounding construction. We show how data is used to analyse the national income, the construction industry and firms within construction. By describing the logic of accounting systems at all levels, it is possible to draw conclusions and gain insights into the operations of firms involved in the production of the built environment. The framework of all accounts is based on stocks and flows and we use this approach to understand construction in Chapter 5. Chapter 6 provides an analysis of the distribution and ownership of the property assets produced by the construction sector in the form of buildings and land.

Part III describes and accounts for the economic roles of the various participants in the construction process. These roles combine in different ways depending on the nature of the project and the type of client. The complexity of the production process in construction is explained further in Chapter 8 on contracting systems. Moreover, many of the conflicts within construction between constructors and their employers can be seen in the fragmentation of the production process described in Chapter 8.

In the fourth and final part of this book, we deal with the wider issues of construction and its relationship to the rest of the economy.

In Chapter 9 we look at the economic concepts of the multiplier and the accelerator in relation to construction and construction firms. Chapter 10 deals with economic cycles and the construction sector and discusses the business cycle and the causes and effects of variations in demand on construction.

Very few books can claim to be the final word on their subject, and *The Economics of the Modern Construction Sector* is no exception. Instead we want the reader to find the contents stimulating and thought provoking. One of the aims of this book is to act as an aid to further research. We hope therefore that many of the questions raised by the book will stimulate research into the economic processes involved in the field of construction economics.

GRAHAM J. IVE
STEPHEN L. GRUNEBERG

List of Abbreviations

ACE	Advanced capitalist economy
APTC	Administrative, Professional, Technical, and Clerical
BS	Building Society
CAD/CAM	Computer aided design and computer aided management
CFR	Construction Forecasting and Research
CIBSE	Chartered Institute of Building Services Engineers
CIC	Construction Industry Council
CIRIA	Construction Industry Research and Information Association
COP	Census of Production
DETR	Department of Environment, Transport and the Regions
DLO	Direct Labour Organisations
DOE	Department of the Environment
EMS	European Monetary System
ENR	Engineering News Record
EU	European Union
FAME	Financial Analysis Made Easy
GCS	Gross Capital Stock
GDFCF	Gross Domestic Fixed Capital Formation
GDP	Gross Domestic Product
GFCF	Gross Fixed Capital Formation
GOP	Gross operating profit
HA	Housing Association
HCS	*Housing and Construction Statistics*
ICE	Institution of Civil Engineers
ICOR	Incremental Capital Output Ratio
ISIC	International Standard Industrial Classification
IT	Information technology
JCT	Joint Contracts Tribunal
LA	Local Authority
LFS	Labour Force Survey
LOSC	Labour only sub-contractor
MBO	Management-buy-out
MES	Minimum efficient scale

NACE (Rev1)	The General Industrial Classification of Economic Activities
NBER	National Bureau of Economic Research
NEDO	National Economic Development Office
NFCF	Net Fixed Capital Formation
NOP	Net operating profit
NPV	Net Present Value
NWRA	National Working Rules Agreement
ONS	Office for National Statistics
O–O	Owner-occupier
PBR	Payment by results
PFI	Private Finance Initiative
R & M	Repair and maintenance
RIBA	Royal Institute of British Architects
ROCE	Return on Capital Employed
RPI	Retail Price Index
SET	Selective Employment Tax
SIC	The Standard Industrial Classification
SPC	Special Purpose Company
SSE	Social structure of the economy
SSHP	Social structure of housing provision
SSOCP	Social systems of organisation of the construction process
SSP	Social structure of production or provision
TGWU	Transport and General Workers Union
UCATT	Union of Construction Trades and Allied Technicians
VIBO	Vertically Integrated Building Owner
WES	Work to existing structures

Introduction

An explanation by way of introduction: the value of realism and the realism of value in economics.

The search for a theory of value in economics is for a 'unifying grand theory' capable, in principle, of explaining the totality of economic phenomena – a search for a common unit of measurement, to be sure, but beyond that, the underlying sufficient cause of values, in the plural (i.e. prices of this and that), and the force capable of yielding a determinate set of economic outcomes from certain non-economic givens – in short, the idea that both back economics' claims to be a science, by giving it a unifying object of study and a method, and at the same time lays it open to accusations of being no more nor less than metaphysics. For value is abstract, and not directly visible.

The great arguments in the history of economics have been arguments between theories of value (Cole *et al.*, 1991; Varoufakis, 1998). Following Cole *et al.* (and many others) we can call these contending theories of value the 'subjective preference', 'cost-of-production' and 'abstract labour' theories. Of these, the academically dominant school of thought, throughout the last century, has been based on the 'subjective preference' theory of value: rational, calculating, self-interested choice between given alternatives.

Economic debate has got lively and deep whenever this dominance has come under challenge – and 'economics' has stagnated into an orthodox body of doctrine whenever that challenge has faded. To be clear: we believe, along with an increasing number of other economists critical of the orthodoxy, that economics took a profound 'wrong turn' when it tied itself, as a 'reputable science', to the subjective preference theory of value in what is known as its 'neo-classical' form. We believe that this mistake has led both to bad theory and a wrong agenda for economics – and explains especially the difficulties which arise when it is attempted to apply economic theory to the interpretation (we will not say, explanation) of the 'real world' of modern industry. Nor, in our view, is it simply a matter of the 'wrong' theory of value having been chosen, whereas the 'right' choice would have solved all problems. Rather, one source of the difficulty for a practical economics lies in the drive, perhaps innate in all value-theories, towards an excess of abstraction, at too great a price in terms of realism and relevance.

However, statements such as those made in the previous paragraph open up such a range of arguments and issues between economists, and moreover arguments necessarily conducted in languages more or less impenetrable to non-economists, that it is certainly not our intention to develop systematically those statements or claims in the work that follows, in the sense of conducting an 'argument' or critique of orthodoxy – for to do so would exclude the possibility of this being a work of utility to students and practitioners in the production of the built environment. Instead, we hope the work will speak for itself, at least in respect of the range of issues our approach enables us to cover, the analytical methods we propose and the agenda of questions (for research and otherwise) which we raise. We do, however, think it both useful and necessary to give 'fair warning', especially to readers who have a knowledge of orthodox neo-classical economics – for such readers will not find in what follows much of what they will expect, and will find much that they will not expect.

We believe that our approach, as exemplified in the chapters that follow, is sufficiently consistent and straightforward as to be reasonably clear *in its application* to economists and non-economists alike. However, we also feel we owe a duty to both kinds of reader to explain in this introduction just what kind of economists we are – that is, what position we take in the fundamental debates that divide economists today.

The approach of many economists when they come to write 'practical' works on the economics of a particular sector or industry, is to start with received value theory, expound a version (to a greater or lesser degree simplified) of that theory, and then to give 'examples' of its application by nominally substituting apparently recognisable phenomena of that industry into the purely formal categories of that theory. An instance would be a discussion of market price for a commodity in terms of the thought-experiment that demonstrates the possibility of a set of pre-reconciled independent choices made by possessors of given productive endowments and given consumption tastes (each with perfect knowledge, and all acting in 'analytic' time to explore all hypothetical options fully before making actual choices) leading to an equilibrium of demand and supply; moves on to describe this imagined market as one of 'perfect competition'; and then discusses how the idea of perfect competition can be used, to some extent or other, to describe and illuminate actual markets for grain, or fish, or stocks and shares.

Our approach, by contrast, is to start by making a model of the actual *processes* by which economic actors arrive at their decisions – in terms which we hope will be recognised by practitioners as capturing certain (inevitably, not all) interesting or significant aspects of that process. We are in a position, since we are concerned here only with the construction industry, to develop and to prefer 'local' or special to general theory – to adapt our models to capture local circumstances, even at the cost of loss of ability to generalise about the economy at large. This we are happy to do.

The substantive content of our theory is to an extent eclectic, formed by taking and melding together, magpie-like, whatever catches our interest from a diverse range of sources. However, just as magpies prefer that which glitters from the array open to them, so we prefer that which looks to us more 'realistic'. Our approach often involves simply relaxing the stays of a pre-existing theoretical corset, by introducing some added degrees of realism.

All of this courts the danger of over-compensating, to the extent that any and all classificatory or descriptive coherence is lost – the point where each instance becomes unique. Naturally, we hope that our readers will agree with us that we have not gone so far as that.

One powerful inspiration for us, in our quest for a sufficient, minimal consistency and theoretical coherence, has been the approach to the treatment of time, uncertainty, surprise, the past and the future. Time, throughout what follows, is we hope almost always *perspective* not *analytic* time. In perspective time, decisions are made sequentially, not simultaneously, and are made using the exercise of economic imagination about possible futures – that is, under real uncertainty, where what is envisioned *ex ante* often differs from what is realised *ex post*. In this, like other economists of construction (Hillebrandt, 1985; Bon, 1989) we have been inspired by the writings of G.L.S. Shackle and other 'Austrian' economists – even where we disagree sharply with the political economy of much of Austrian economics. It is to Shackle, however, that we owe a particular debt – and Shackle was both an 'Austrian' and a post-Keynesian.

The post-Keynesian economists are, of course, best known for their macroeconomics – and for insisting upon the profound theoretical implications of Keynes' critique of macroeconomic orthodoxy, and thus resisting the absorption of the Keynesian legacy into a slightly modified orthodoxy – the so-called Keynesian neo-classical synthesis. However, there is also a very distinctive and coherent post-Keynesian microeconomics, and this has often provided us with at least a starting point or point of reference. We are also indebted to those economists who have sought to fuse elements of Keynesian and Marxian economics – the tradition beginning with Kalecki and Robinson, and carried on today by, *inter alia*, such inspirations for parts of our work as Bowles, Weisskopf and Marglin.

Nearly twenty years ago, a book appeared whose publication in our view (and that of many) deserves to be seen as seminal to the development of a truly modern, non-orthodox economics – Nelson and Winter's *An Evolutionary Theory of Economic Change* (1982). The 'Introduction' to that work contains an argument about the nature of the malaise afflicting orthodox microeconomic theory (what is known to economists as general equilibrium theory), and directions for its remedy, with which we wholly agree. Indeed we could wish, if academic convention and copyright law did not forbid us, to reproduce it more or less entire as our own introduction to the present work. Among the very many indications therein for positive directions for a new microeconomics let us cite at least the following:

(1) theory must seek to comprehend, in stylised settings, the unfolding of economic events over (perspective) time; we must escape from the grasp of a purely 'analytic' time, that is really no time at all.
(2) firms are motivated by profit, seek it and search for ways to increase it, but 'firms' actions will not be assumed to be profit-maximising over well-defined and exogenously given choice sets' (p. 4).
(3) analysis should not focus on hypothetical states of industry equilibrium, in which all unprofitable firms have left the industry and the profitable ones are at their desired size.
(4) firms learn; at any given time a firm has limited capabilities, and habitually uses certain decision-rules: 'Over time these ... are modified as a result of both deliberate problem solving efforts and random events.'

Meanwhile, orthodoxy itself has been invaded by the practitioners of the so-called 'new institutional economics', an economics of incomplete information, bounded rationality, complex organisations and transaction and organisational costs. This we welcome, and use happily where the only developed alternative is the old orthodoxy – while recognising that it too is suffering the fate of Keynesian economics – to be re-absorbed, in somewhat travestied form, into a revised orthodoxy.

This, then, is the kind of economics, and the approach to the use of economics to study industry and business, that a reader will find exemplified in what follows. We hope you, the reader, will find it yields both light and fruit.

Part I
Production

1
Construction and Value Added

Introduction

Production is continually changing. Output is always growing or declining and altering in composition. Products and services are made in different ways. The uses to which output is put change. The size and number of providers varies over time. It is therefore difficult to define the limits of the construction sector without making an arbitrary cut off at some point. This would apply to any branch of production, not just construction. Nevertheless, it is clear both theoretically and from published statistics that the construction sector is an important, large, distinct, though not isolated, part of the economy.

In 1991, out of a total UK population of 57 million people, 24 million were engaged in paid employment. Of the 24 million in work, approximately 5 per cent, or around 1.2 million, were employed directly in construction on building and civil engineering projects by construction firms. This does not include the self-employed in the construction industry, nor those employed by manufacturers of building components and the suppliers of materials such as aggregates, cement and bricks. Nor does the figure include professional architects, surveyors and construction engineers and their office support staff. Many other people also work casually, for part of the year, in the building sector. Finally, a great many people carry out work for themselves, improving, maintaining or participating in self build schemes. The work of all of the above constitutes the process of production of the built environment. This chapter defines construction and shows how productive processes involve adding value to inputs.

Defining the construction sector of production

The production of the built environment includes any and all of the activities which contribute to the creation of a certain kind of object, namely buildings and other fixed structures, such as bridges and dams. Many of these activities take place before the materials and components arrive at the site. Work *in situ* is only the last stage of the production process and, by convention, is said to define the *construction* part of the production of the built environment. *In situ* workers are defined as the work force of the *construction industry*.

Construction is a certain kind of production of a certain kind of object with a certain kind of use. All construction industry output shares a common characteristic. It is a product, regardless of use, that is fixed in place to a site. Both buildings and some civil engineering structures, such as reservoirs, provide *bounded* spaces. Generally, buildings shelter and accommodate specific activities, and may therefore be defined by reference to those activities. Dwellings provide spaces for residential activities, while commercial activities take place within commercial buildings. *Infrastructure*, on the other hand links bounded spaces and facilitates movement, distribution, or transfer *through* space. Goods and people are enabled to move by the infrastructure provided by road, rail and other transport systems. Electric power, gas and water, for instance, are distributed through infrastructure systems and information is transferred through space through the infrastructure of telecommunication systems.

Production includes various activities from conception through design to execution. Design includes design of components as well as final products. Execution involves the preparation of the site, movement of materials, working-up of materials, making of sub-assemblies and final assembly.

The objects of the built environment include buildings and infrastructure, which in practice are often discussed in terms of their construction elements, such as sub-structure, superstructure, finishes and services. In general we might also distinguish *replacement* of buildings from *additional buildings*, and *alteration* from *maintenance* of existing buildings.

We will use this framework of production, product and use, to answer questions like, who consumes buildings?; how are buildings produced?; when are buildings produced and why? These and other questions can only be answered in relation to a particular mode of social organisation. Let us suppose for instance, that we wish to describe only societies with the following defining characteristics:

(a) property rights in land, so that we can have the category *landowners*
(b) exchange of buildings as commodities, so that we can have *producers, consumers* and *merchants*
(c) borrowing and lending of money at interest, so that we can have *financiers*
(d) alienation rights in land, so that we can have *buyers* and *sellers* of land
(e) a horizontal division of labour, so that we have *building producers* who are in separate building *trades* or occupations
(f) property rights in buildings, separable from rights in the land on which they stand, so that at any one time there may be separate landowners and *building owners*
(g) commodification of labour power, so that we can have *employers* and *employees*
(h) separation of design from execution, and, more generally, of mental and manual labour, so that we have *designers* and *constructors*
(i) production directed by and in large organisations, especially firms, involving *managers* as well as *owners*.

These social conditions embrace an enormous range of specific societies, which are nevertheless each different in important respects. For example, these characteristics would apply to Britain at every period since the sixteenth century but that does not mean to say there have not been many substantial changes even within the last forty years in the forms, roles and actors constituting the construction process in the United Kingdom. We shall seek to demonstrate these changes in later sections of this book. It is within the terms of this set of social relations that it makes sense to think in terms of *consumers* of buildings and to relate these consumers to *producers* and producers to one another in a particular way. These social relations provide the context for an understanding of the economics of the modern UK construction sector.

We define the **construction sector** as all production activities contributing to the **production of the built environment**. In other words, we first define the final product, the built environment, and then group together activities contributing at each stage to the transformation of natural resources into that final product.

The matrix in Figure 1.1 shows the extent of the built environment sector, the food sector and the car sector in terms of the final goods and services consumers or households purchase. The final demand by consumers is for finished goods and services, which they use

Table 1.1 Final demand products and sectors by activities

	Built environment	Food	Cars
Primary sector	Quarrying, mining	Farming	Mining
Secondary sector	Material and component manufacturing, construction	Food processing and packing	Component manufacturing, assembly
Tertiary sector	Plant hire, professional services, property letting	Catering, retailing, food distributing	Car dealing, car hiring, car servicing

themselves. Consumers do not in general buy goods with a view to selling them on at a profit.

Table 1.1 shows industries grouped into three broad categories, primary, secondary and tertiary, reflecting different stages of production. The primary sector is concerned with the direct appropriation of earth bound natural resources for economic purposes and includes agriculture, forestry, fishing, mining and quarrying.

In the secondary sector industries use the outputs of the primary sector and convert or transform them in a series of operations until they are ready for use by final users. The secondary sector includes all manufacturing, energy and water industries, as well as construction. The products of these first two sectors are called goods.

The products of the tertiary sector are called services. It is more mixed in character than the first two sectors. The service sector includes the general physical distribution of goods, from their transportation out of their place of production to their place of sale or consumption. As part of the distribution process, this sector includes the wholesaling and retailing of goods. Other services included in the tertiary sector of the economy include services provided to consumers, such as live music, restaurant meals and hairdressing, and unlike the tangible goods produced in the primary and secondary sectors, these services cannot be stored. It also includes provision of financial services, such as banking. Financial services are essentially concerned with financial assets and the legal claims to ownership of physical goods and services.

In addition, this third sector of the economy includes the provision of business or professional services, such as law and accountancy, to other firms in all sectors. Finally this sector includes all public services such as health, education, the military, police, judiciary, etc. Essentially, the tertiary sector consists of the direct production of services, plus all the commercial activities of buying and selling commodities without producing new commodities, plus the financial activities of banking institutions, of deposit taking and lending which create financial assets and liabilities.

As shown in Figure 1.1, the construction industry is part of the process of producing and maintaining the built environment and may be classed in the secondary sector because it transforms manufactured materials into final products. However, the built environment production sector also consists of quarries, manufacture of construction materials and components, and provision of related professional services such as design, engineering and cost control. Many other types of firm are also directly involved from plant hire firms, to estate agents, developers and property companies. The process of producing the built environment spans the primary, secondary and tertiary sectors.

In practice, it is not possible to define the construction *industry* to make it synonymous with the execution of all construction activity. The basic way any industry is defined is as a set of firms. The population of firms in the economy is divided into industries on the principle of *potential competition*. Firms are put in the same industry if they produce outputs which are similar or reasonably close substitutes for one another, or if they use similar technology and materials, and are therefore in competition in the markets to buy these inputs. If these two criteria conflict, it is the similarity of inputs that is the more important criterion used to define an industry, whereas similarity of outputs is used to define a *market*.

The production of steel and the manufacture of furniture are obviously regarded as separate industries. However, within an industry there can be several different markets. From the perspective of markets for their products, furniture for domestic use might be regarded as one market, with office furniture another market; whereas from the perspective of industries, manufacture of steel furniture might be regarded as one industry, and manufacture of wooden furniture as another.

The construction sector certainly comprises several industries and several markets. Its constituent industries comprise sets of firms engaged in each stage of the process of production of the built environment. Thus the firms of each stage compete directly (actually or potentially) with one another, and thus constitute an industry. The firms of other stages in the process stand not as competitors but as suppliers to or buyers from that industry, and therefore are part of other industries. Many firms involved in the construction process are actually involved at several different stages. Sometimes this is true of a firm's involvement in a single building project so that the same firm, for example, makes and supplies roadstone and undertakes the road building contract for the same project. Often, the firm has a range of construction related businesses each operating separately, so that they are not even normally found working together on the same projects. Some construction firms, though not many, also have interests outside the construction sector altogether.

Now, whilst the brick making industry, for example, clearly falls within our definition of the construction sector, as virtually all of its output is used as intermediate input into the production of the built environment, the same is not true of, say, the glass or steel industries. Much of their output, perhaps the largest part of it, is indeed used in buildings, but much of it has other final uses – in the making of automobiles, for example. Firms in these industries *belong* to several final goods or end use sectors, not to one. Accurate figures for the proportion of their output devoted to construction are not available on a firm-by-firm basis, though this information is available, for example from Input–Output Tables (p. 19), for each of these industries as a whole.

In principle, we could set about to identify and define as many *industries* within construction as we believe to be appropriate. For example, we would certainly wish to recognise a built environment design industry, comprising firms of architects, civil, structural and building services engineers. However, to be of practical use we need to be able to measure and obtain data about an industry, and unfortunately, we are often at the mercy of the imperfections of our data sources. The Standard Industrial Classification (SIC), our main official source, recognises no such industry as built environment design. Instead, these activities are grouped with many others having no direct relation to the construction process, under the heading of 'Architectural and engineering activities and related technical consultancy'. Fortunately, the SIC is by no means our only data source. Most industries have trade associations, which publish data collected from their member firms. There are also trade magazines. Data about parts of the built environment design industry, for instance, is available from the RIBA, ICE, and CIBSE, as well as in the pages of *ENR*, *Building* and other journals. Recent work by Construction Forecasting Research (CFR) for the Construction Industry Council (CIC) will yield regular statistics on the built environment design industry for the first time.

Nevertheless the SIC is a useful starting point for gathering statistics on any industry. Each establishment of a firm is a statistical unit, which must be allocated to a particular industry, usually depending on the *main* activity of the establishment, which depends on the main type of output produced. In this way, in principle, diverse statistics on output, employment and prices can all be related to the same firms, industry or product. Because new processes and products develop over time, there is a need to revise the definitions of industries to classify economic activities in their most appropriate categories. The SIC 80 replaced the SIC 68 of 1968, but in turn it has now been replaced by the SIC 92.

The SIC 92 is the current classification of UK firms used in government statistics and is consistent with the industrial categories used across the European Union. The General Industrial Classification of Economic Activities within the European Communities, otherwise known as NACE (Rev1), allows statistics from the different member states to be compared. In 1989, for the same reason of international comparability of industrial statistics, the United Nations agreed an International Standard Industrial Classification of All Economic Activities, (ISIC Rev 3).

In the SIC 92, industries are divided into 17 sections from A to Q, and each section is further broken down into groups, which contain classes and subclasses of firms. Construction industry is section F while real estate is a group within section K. The SIC 92 classification of firms involved in the construction sector is given in the appendix to this chapter. In the appendix there are two lists. The construction sector comprises the whole output of industries in List A and a large proportion of the output of industries in List B.

A *narrow* definition of the construction industry includes only those firms undertaking on-site assembly. Nevertheless, this is the most important single part of the construction sector, in terms of its share or contribution to the total value of the production of the built environment. It has sometimes been argued that even this narrow definition of the construction industry should be regarded as several separate industries, and, indeed, the NACE classification attempts to do this.

The SIC 92 attempts to divide construction into various categories such as:

45.21 General construction of buildings and civil engineering works
45.23 Construction of highways, roads, airfields and sport facilities

45.31 Installation of electrical wiring and fittings
45.33 Plumbing
45.42 Joinery installation
45.43 Floor and wall covering
45.44 Painting and glazing
45.45 Other building completion.

However, these sub-divisions are rather unsatisfactory. The first category of firms, for example, refers to those companies engaged in general construction of buildings and civil engineering works but this category is composed of a miscellany of firms whose work covers 'all types of buildings'. Moreover, the term 'general construction of buildings and civil engineering works' also includes civil engineering constructions such as bridges, elevated highways and tunnels and clearly overlaps with the second category of firms engaged in the construction of highways, roads, airfields and sport facilities.

In any case the Department of the Environment, Transport and the Regions (DETR), one of the government departments charged with responsibility for gathering statistics on the construction industry, does not use these SIC categories in its own surveys of GB firms. The *traditional* British approach was to divide the construction industry first into main and subcontractors and then divide main contractors into builders and civil engineers. Subcontractors were divided into a long list of *trades*. The DETR uses a classification very similar to that, except that it introduces an element of confusion by referring to *main trades* and *specialist trades*, as if all subcontractors worked in *specialist* trades, and as if main contractors themselves still undertook work in certain *main* or basic trades.

Figure 1.1 illustrates the relationship between the main industries involved in the production of the built environment. One might call this a *broad* definition of the construction sector as it includes many firms in industries from List B in the Appendix.

In 1980, according to Janssen (1983), the production of the built environment was equivalent to 18.5 per cent of the West German national income. If this proportion were reflected in the percentage of the working population of West Germany in 1980, then approximately 4.6 million people would have been engaged in production of the built environment. Of the 4.6 million people, it was estimated that only 28 per cent or 1.3 million were involved as management or labour in the building assembly process on site.

Using a combination of data published in *Economic Trends* (1994) and the *Employment Gazette* (1995), it is possible to estimate the

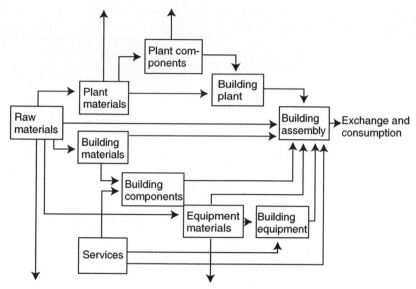

Figure 1.1 Main sectors of production and service contributing to building production
Source: Based on Janssen (1983).

number of people engaged directly and indirectly in the production of the built environment in the UK. Unfortunately the two sources of data use different industrial classifications, the former based on 123 industrial groupings and the latter on the SIC 1980 making the figures not strictly equivalent. Nevertheless the result shown in Table 1.2 is an approximation which indicates the level of employment generated by construction activity.

From Table 1.2 almost 2.5 million people out of a workforce of just over 22.3 million, over 11.2 per cent of employees in employment are engaged in the production of the built environment. This compares with the figure of 4.7 per cent taking the construction industry category alone.

Is the construction industry really several different industries?

The argument that construction is not one but several industries rests on two propositions. The first proposition is that most *firms* are actu-

Table 1.2   Estimate of people engaged directly and indirectly in the production of the built environment, 1993

Industrial sector	Output % going to construction	Employment	Labour equivalent engaged on construction related output.
Agriculture	0.02	281,000	66
Energy	1.65	443,000	7,302
Manufacturing	13.12	5,012,000	657,805
Construction	**100.00**	**1,060,000**	**1,060,000**
Distribution	10.43	3,551,000	370,372
Transport	2.09	925,000	19,298
Business services	6.77	3,135,000	212,312
Other services	2.50	6,748,000	168,437
Other sectors	0.0	1,198,000	0
Total		**22,353,000**	**2,495,592**

Sources: *Economic Trends* (1994), Table 2, 1990 domestic use matrix (London: HMSO).
Employment Gazette (1995), Table 1.2, Employees in employment in Great Britain in 1990, seasonally adjusted (London: HMSO).

ally specialised in building or civil engineering but not in both. However, Ball (1988) and Ive (1983) both show that parent companies commonly own different subsidiary or operating companies within so called sub-industries. Thus almost all the major firms in construction have separate divisions operating in civil engineering, building contracting and speculative house building. They move capital, and to a degree staff, around and between them. At the parent or group level, therefore, it is mostly *not* the case that the larger firms are specialised in either one or another of these activities.

The second proposition is that the *resources* used in, say, civil engineering are largely different from those used in building, and are therefore not even potentially transferable between the two. As a result firms in these two sub-industries would not be in competition for the same resources. However, NEDO (1978) showed that in many respects construction resources, especially labour, are to a degree *flexible* between sub-industries.

Thus Ball, Ive and NEDO all hold to the concept of a unified construction industry. Ball (1988) points out that though firms in the construction industry compete with each other directly for the same customers to a greater or lesser extent, the same could be said of firms in any recognised industry. In other words, the fact that the same firms do not compete with each other for all their work all the time signifies only that there are separate markets within construction. These markets, however, are not the same as sub-industries.

However the opposite can also be argued. Namely, the construction industry is not a single industry but is composed of a number of separate sub-industries. Plant, skills and materials used by civil engineering firms are not the same as those used in building firms. Skills and machinery also vary between different types of specialist firms, although there is a great deal of overlap in the kinds of resources firms use. However, when we define types of resource very narrowly so that there is no overlap, and resources are truly specialised, then the number of specialisations becomes very high – too many for each to be treated as a separate *industry*. The many separate specialist firms simply supply an element in the building process within a highly fragmented construction industry.

Below, we shall develop the view that it makes sense to regard construction as fundamentally split into just three sub-industries. These three sub-industries are main contracting of all kinds, subcontracting, and speculative building. Each of these has certain fundamental and distinctive *business characteristics* that make the differences between them outweigh the differences within them.

Factors of production

Every industrial process is concerned with transforming inputs into outputs using the resources at the command of each firm. These resources are land, labour and capital and are known collectively as *the factors of production*. That is, annual output depends on the input quantities of these productive factors in the same year. The quantity produced reflects the productive powers and capacity of present human knowledge, skill and effort (i.e. labour), the use of accumulated stocks of means of production, including materials, plant and machinery, produced in the past and not consumed (i.e. capital), and the use of stock of natural resources (i.e. land).

This causes several important problems. First, it means we need to distinguish between the amount of a factor existing at any year, which we might be able to measure, and the amount of input or use of it during that year, which is much harder to measure. In the case of capital, for instance, we need to distinguish between the amount of capital stock existing (and therefore owned by someone) and the amount of use made of that stock. If, for example, factories move to working their machines for more or less hours per year than previously, there will be a change in the quantity of the capital factor used, and hence a corresponding change in the amount of output, without any change in the amount of capital in existence.

Second, these factor inputs actually exist in diverse, heterogeneous forms. We can say that there exist so many carpenters and so many steel erectors and so many architects, for example, and so many machines of each actual type – but adding them together into a composite quantity of a factor input is difficult. Do all workers, for example, count as one equal unit of the factor 'labour', regardless of degree of skill, knowledge and training? Can we reduce all machines to quantities of a common physical unit such as horse power?

In practice, the quantity of labour is most often measured in a common physical unit, such as the number of workers, or the number of labour-hours of input used. However, using physical units to quantify inputs ignores quality differences between inputs.

A broad definition of physical capital includes all assets which are (or can be converted into) the means of production. It consists of buildings, infrastructure plant and equipment, and stocks of materials. Wealth, by contrast, refers to these but also to any real asset, including domestic housing, private cars or any other consumer durable, which can be sold or used as collateral for financing production. The value of these assets depends on their second hand price. A more narrow definition of capital as a factor of production refers only to the buildings, plant and machinery, and the stocks of materials and goods actually used in the production process. The term 'capital' is also used, confusingly, to refer to the sum of money, and other financial assets. These, however, are not themselves factors of production, but only give their owners claims in or over physical capital and output.

Inputs of the factor capital are measured in monetary units. However, using monetary units introduces a circularity of reasoning. The monetary value of a piece of capital, such as a building or a machine, is most usually *derived* from an estimate of the future profit or rent income that it is expected to yield for its owner. The value is a capitalisation into a single lump sum, of the expected stream of income receipts over its remaining economic life. But, this then means that we are not measuring factor inputs independently of measuring incomes. If income changes (say, if actual or expected profit income increases) then there will *appear* to have been a corresponding change in the quantity of the input of capital, as firms re-evaluate their assets, especially their property. When there is an actual increase, say, in profit income, we may wish to know whether this is because of an increase in total capital input, which is an increase in the size of the firm, or because of an increase in profit per unit of capital input, perhaps because of an increase in the use of existing plant and machinery.

Alternatively, the value of a piece of capital equipment may be measured by its original purchase price or historic cost, adjusted for depreciation. If the historic purchase price reflects its cost of production, and if the cost of production reflects the quantity of factor inputs used to produce the item, then historic cost may be a tolerable measure of the quantity of factor inputs embodied in that piece of equipment. Only under those conditions can historic cost be used as a measure of the quantity of capital represented by that particular item.

Third, the concept of capital as a factor of production introduces a serious potential confusion between different meanings of the term 'capital'. The term is used in its ordinary, business sense, to mean the ownership of a business, and therefore the ownership of its revenues, and of whatever net revenues or profits are left after all costs have been met out of those revenues. Thus capital means ownership of the entirety of a business, and gives its owners income in the form of profit. However, capital as a factor of production is something else and has several meanings. For instance, consider fixed capital. Fixed capital consists of the physical stocks of machines and buildings used in production, and the contribution their use makes to output, in the sense that the more of these machines there are, the greater can be the output. This is capital as the means of production, and as a distinct 'factor' of production. This kind of capital is then said to earn a return, and this is where the heart of the confusion lies. Keynes called this return the 'efficiency of capital', in an attempt to capture the idea of contribution of capital use to output.

However, it is crucial not to confuse this 'return' of capital defined as a means of production, with the concept of profit as a return on money capital owned. The two would only be identical *if* the owners of a business always captured a share of the output revenues of that business that was just equal to the contribution, which the means of production made to that output. In other words, it is perfectly possible in a modern company for the means of production to generate revenues which the owners, or shareholders, do not receive. Now, it is possible to construct a theory of output and income distribution on a set of assumptions such that this identity would, theoretically, occur. This theory, or rather, this set of axioms, is known as the 'neoclassical theory of value and distribution', and it is in fact offered as 'the' explanation of profit, interest and wages in most current economics textbooks. As will become clear in later chapters, it is not the theoretical model for this work.

Capital is in fact the most complicated, many-faceted and difficult concept in the whole of economics. This is not surprising when we reflect that economics is really all about understanding how a *capitalist* economy works. Naturally, the understanding of capital will be at the core of this whole scientific project.

Land as a factor of production includes the raw materials, such as minerals and oil, contained within it as well as its natural ability to grow food produce. The greater the quantity and the better the quality of natural resources used in production then, other things being equal, the greater will be the output of an economy. In construction, the assembly of buildings on site involves land, but this land is usually owned by the client, rather than the producer. Unlike other forms of production in which the producer must pay for the land on which production takes place, contractors only pay for the land used by them as their head offices and permanent storage spaces. They do not need to pay for the land where production takes place.

The role of *labour* as a factor of production is concerned with the direct and current human input into the production process. The stock of physical capital can be usefully regarded as the result of indirect, past human input or activity. Labour is the contribution of work and effort necessary for production to take place, and includes both the physical and mental activities of skilled, semi-skilled and unskilled workers as well as their managers. It does not, however, include the activities of owners or shareholders.

Incomes are derived from the sale of one's labour, the ownership of capital or the ownership of landed property. Labour incomes are wages. Incomes deriving from the ownership of capital are profits, including interest, and incomes from the ownership of landed property are called rents. The source of these incomes comes from the ability of firms to add value to material inputs through a production process producing and then selling the output with a value greater than that of the inputs used up in its production.

The structure of value added in the production of the built environment

Table 1.3 is a model showing both the process of adding value at each stage in the production process and the distribution of income. To understand Table 1.3, the process begins with the manufacturers of basic materials who sell their output to component manufacturers for £20,000. We assume the manufacturers of basic materials do not buy in any raw materials and only have to pay wages. The component manufacturers sell their output for £30,000 but must deduct the cost of £20,000 for their inputs in order to calculate their income. The difference between the value of sales and the cost of bought-in materials is the value added to the inputs by the component manufacturers. In this case, the value added is £10,000 and this is, in fact, the source of income of the owners and workers in each firm.

It should be noted that although the total value of sales receipts is £130,000, total income is only £45,000. The reason for this difference is because the value of sales receipts of each firm always includes the value of each firm's inputs and the total of all receipts therefore counts the value of inputs several times, an error economists call double counting. The value added approach prevents double counting because it only calculates the *contribution* to its inputs made by each firm and not the inputs themselves.

From Table 1.3 it would appear that the construction industry contributed 7.19 per cent of the GDP in 1989 and 5.13 per cent in 1997. However, as noted above, this figure does not take the manufacture of building components into account nor the value of the professional services provided by architects, surveyors and others. Nevertheless

Table 1.3　Value added by industry, 1989 and 1997 at current basic prices

Industry	Value added (£m) in 1989	Percentage contribution to GDP in 1989	Value added (£m) in 1997	Percentage contribution to GDP in 1997
Agriculture	9,097	1.98	10,820	1.52
Mining and quarrying	12,491	2.71	18,137	2.55
Manufacturing	110,407	23.97	146,522	20.60
Electricity, gas and water	11,514	2.50	16,227	2.28
Construction	33,117	7.19	36,491	5.13
Wholesale and retail trade and (non-construction) repair	52,423	11.38	83,316	11.71
Transport and communication	38,818	8.43	59,694	8.39
Business services and property	75,328	16.36	140,078	19.69
Public administration and defence	29,514	6.41	38,940	5.47
Education, health and social work	48,238	10.47	85,129	11.97
Other services	39,589	8.60	75,916	10.67
GDP (all industries)	460,536	100	711,270	1000

Source: United Kingdom National Accounts, 1998 edition, ONS, HMSO.

£36,491m is a measure of the value added by the narrowly defined construction industry towards the total value of the GDP in 1997.

Of course Table 1.3 describes only one channel of production, a channel being the progression of a raw material input through to a finished product or service. Each channel therefore contributes a proportion of the value of the final goods or services. This model does not include all the channels of production which would be required to produce a completed building. For instance, there is a similar channel of production beginning, say, with basic materials manufacturers who produce the material inputs and components to the manufacturers of construction plant, who then supply plant hirers, who in turn provide services to subcontractors, whose clients are the main contractors.

All production is similarly concerned with transforming inputs into outputs. Primary sector firms which extract raw materials are in basic industries, such as mining and quarrying. These firms then pass their output of raw materials on to firms in sub-basic industries, such as the steel industry. These are classed in the secondary sector. In sub-basic industries raw materials are processed into usable inputs for manufacturers in the rest of the secondary sector. Construction takes place in the secondary sector by transforming inputs of materials or components into finished structures by assembling components and materials on site.

Construction, despite claims to the contrary, remains an industry whose physicality is very high – i.e. one for which the value of material inputs and labour input is very large as a percentage of output value. A smaller number of architects or engineers manage to command fees which bear no close relation to their costs of material, labour and fixed capital inputs.

Equation 1.1 is a technical input–output ratio showing the quantity of output derived from a quantity of inputs. Thus

Technical input–output ratio = O_q/I_q (1.1)
where

O_q = quantity of outputs
I_q = quantity of inputs.

This is a physical and technical relationship. Thus, an input–output ratio of, say, seed corn to harvested corn might be 1:10. The inverse of the input–output ratio is the input–output coefficient and in this case it is equal to 10. This coefficient can be used to predict the output from a given input. In this instance, it shows that for every one unit of input of corn, 10 units are harvested. This technical concept of input–output ratios can be taken further by introducing money values in the form of prices. Thus;

$$(I_p \times I_q) \rightarrow (O_p \times O_q) \tag{1.2}$$

where

I_p = price of inputs
O_p = price of outputs

In (1.2) production transforms the quantity of input multiplied by its price into the quantity of output multiplied by its selling price. If the technical ratio, O_q/I_q, and the ratio of output prices to input prices, O_p/I_p, remain the same, so will the input-output coefficient. Otherwise, the coefficient will change in response to changes in technology or changes in relative prices.

The value added is the difference between the value of inputs (excluding immediate labour and fixed capital, which are not treated as inputs) and the value of outputs. During the production process materials and components may move from one firm to the next as each firm uses another firm's output as its inputs and transforms that input into the next stage in the production of a finished product. In construction, finished products are classed as capital goods, whether they are housing, offices and factories or infrastructure, such as roads. As a final product the output is used, maintained and consumed by the buyer. For this reason it can be said to derive its price partly from the *value* of its use placed on it by the final purchaser, in the sense of willingness to pay a sum of money to obtain the benefits of that use. The price of a final product is therefore normally equal to or less than its use value to the final purchaser.

Input–output analysis and the value added by the construction sector

Every industry processes inputs supplied by other industries or sectors of the economy, mainly materials and manufactured components. One of the uses of input-output analysis is to understand the interdependency between the construction industry and its suppliers as well as the industries which it in turn supplies.

As we have noted, economists divide production into production of intermediate and final outputs. Intermediate outputs are those goods and services purchased by other firms as inputs to their production of further commodities. Its stock of intermediate goods forms part of a firm's circulating capital. All chains of production of intermediate inputs must ultimately end in the production of final outputs. For instance, manufactured building components are intermediate goods which are then assembled on site to produce a completed building which is the final output.

Final outputs include those goods and services purchased by households and government agencies for consumption. Fixed capital goods, including buildings, are also deemed to be final goods, because (with the exception of property speculators), they do not become part of another firm's short run circulating capital. However, unlike consumption goods, fixed capital goods have not finally left the sphere of circulation of capital, though they are not part of the *current* circulation. Current circulation refers to production for sale within the current time period. The fixed capital goods, when installed and in use, contribute to capacity to produce and sell commodities in current and future time periods.

Purchases of final outputs are described as final demand. Demand for intermediate outputs is a derived demand, that is, it depends upon the level of final demand. Every industry produces a combination of both intermediate goods for sale to other firms and final goods for sale to final users.

The output of each industry may therefore be calculated by summing the value of its production of final and intermediate goods. Intermediate goods are the inputs to other industries. To explore these relationships, input-output analysis was developed by Wassily Leontief in the 1930s. Input-output analysis measures the value of inputs of each industry which come from other industries. The distinction between one industry and another is in practice not clear cut. Firms in any industry tend to produce a principal product as well as secondary outputs which ideally would come under another industry. Thus the problem of defining industries by firms means that some of the output is registered as a separate commodity, while some of the output of a particular commodity is produced by firms in another industry.

Nevertheless, it is possible to use input-output tables to estimate the value added to inputs by industries. The value added by each industry is its contribution to the national income. According to the United Kingdom National Accounts, in 1989, a peak year for construction, and in 1997, the most recent year for which data were available, the contributions to national income by different industries are shown in Table 1.4.

The value added approach measures the value added by the construction industry to its inputs from other sectors. It is this added value which forms the source of income of those who own firms or work in the industry. The question then arises as to how the value added is shared between the owners of the firms and their employees. This is not only a question of industrial relations but also an issue related to the concepts of the economic surplus and of the net product of economic activity.

From the point of view of the economy as a whole, value added might be thought to be the objective or purpose of production, and the target to be maximised. The higher the sum of value added, which is the GDP, the higher is the net increase in goods and services over and above what would have existed if production had not taken place. This increase in the stock of goods and services available to be used is the net product of economic activity, after deducting from the gross product of a period, all prior stocks that have been used up as a result of production in that period. These have to be deducted both in order to see the effect of present production on income and wealth, and to see its effect on future capacity to produce.

From Table 1.4 the construction industry contributed 7.2 per cent of the GDP in 1989 and 5.9 per cent in 1997. However, as noted above, this

Table 1.4   Value added by industry 1989

Industry	Value added (£m) in 1989	Percentage contribution to GDP in 1989	Value added (£m) in 1997	Percentage contribution to GDP in 1997
Agriculture	9,097	2.0	10,820	1.5
Mining and quarrying	12,491	2.7	18,137	2.6
Manufacturing	110,407	24.0	146,522	20.6
Electricity, gas and water	11,514	2.5	16,227	2.3
Construction	33,117	7.2	36,491	5.1
Wholesale and retail trade and (non-construction) repair	52,423	11.4	83,316	11.7
Transport and communication	38,818	8.4	59,694	8.4
Business services and property	75,328	16.4	140,078	19.7
Public administration and defence	29,514	6.4	38,940	5.5
Education, health and social work	48,238	10.5	85,129	12.0
Other services	39,589	8.6	75,916	10.7
GDP (all industries)	460,536	100	711,270	100

Source: United Kingdom National Accounts, 1998 edn, ONS (London: HMSO).

figure does not take the manufacture of building components into account nor the value of the professional services provided by architects, surveyors and others. Nevertheless £36,491 million is a measure of the value added by the narrowly defined construction industry towards the total value of the GDP in 1997.

If used up stocks are not replaced, then the future productive capacity of the economy will be correspondingly reduced, producing less gross output in future periods. Likewise, if the money sums necessary to pay for this replacement are not deducted from gross income, we will believe that our income is more than is really the case. These prior stocks were part of wealth. Proceeds from sale of gross output appear to be income, and therefore available to be spent on consumption. But if we collectively consume the gross output, we will find that our collective wealth is now less than before. In effect, we have eaten the seed corn – transferred a stock of wealth into a flow of current income and consumption.

The net product, then, is the measure of the *benefit* from production. In one sense it is the surplus of what exists after production over what existed before production. However, not only do stocks of goods have to be kept intact but so do stocks (populations) of people. Suppose we regard this year's *necessary* consumption to be equal to last year's actual consumption because of the ratchet effect establishing expected and customary standards of living, we can then divide the net product into the part *necessary* to maintain that standard of consumption and the balance, the *surplus product*. This surplus or increment is then shared between increased wages and gross profits.

If wages were constant, then the whole of this surplus product would accrue to owners of firms as their gross profit. However, wages are not constant. Moreover, the workers in a firm may pitch the wage demands by looking at the trend in and size of the surplus generated by their firm's production. The larger the surplus, the higher the wage demands. If this is the case, we have an explicit fight between owners and workers over the division of the surplus between profits and increased wages. In practice, things become more complicated if firms are able to respond to wage increases by raising prices and revenues so as to maintain the value of surplus appropriated as gross profit.

The concept of the *economic surplus* is a macro-economic one. It has its reality at the level of the economy as a whole, and the aggregate division of GDP into profits and wages. By contrast, the concept of gross profit is a micro-economic concept reflecting the reality as experienced by the owners and managers of a single firm.

Gross and net profit

A profit arises for a firm after all production costs have been met. From the point of view of the owners of a firm, production costs include materials, labour, energy, as well as the cost of replacement when plant and machinery wears down. Production costs do not include rent, interest and profit or taxes. If sales revenues only covered production costs there would be no funds or surplus available for rent, interest or net profit. The size of the gross profit therefore depends on the excess of revenues over production costs.

Materials are bought in competition with other firms in the same industry as well as other firms in other industries. Glass, for instance, may be purchased by firms in construction, car production and ship building. The greater the cost of materials the higher the production costs and the lower is the gross profit, unless the firm can pass on the higher cost in higher prices to increase its revenue.

Labour is the major cost facing most construction firms. Just as with the cost of materials, the greater the amount paid to labour, the less will be the gross profit. There is therefore a conflict of interest between labour and the receivers of profit, interest and rent. Net profit of a firm is its gross profit less its payments of interest and rent.

Net profit can be used to re-invest in a firm for it to expand its production by increasing its output, diversifying into new markets or by research and development of new products. Such re-investment is mostly funded from retained net profits and is essential for the survival of firms. The alternative use of net profits is to reward shareholders by distributing a proportion of the profits in the form of dividend payments. Without this reward shareholders would not have an incentive to continue holding shares in a firm. These issues will be dealt with later in the book.

Concluding remarks

By looking at construction as the total production process involving the transformation of raw materials into finished buildings and structures, the sector can be seen as much larger, more complex, and far more significant economically than much official data would suggest. This is because official data is usually based on the SIC, which in turn is based on materials and individual processes. This means that important production processes within the construction sector are excluded from construction industry statistics, rather than looking at the resources used altogether in the provision of the built environment.

As with any production process the construction industry can be seen as a series of stages. Each stage consists of inputs, starting with the raw materials extracted from the land, which are then worked on to produce outputs. These outputs become the inputs of the next stage and so on until the final product is sold. In this way each stage makes a contribution to the eventual final product by adding value to the inputs it buys from the previous stage. Value added at any stage is the source of both profits and wages.

Appendix: Standard Industrial Classification, 1992

Appendix Table 1A   Summary of SIC 92 definition of construction and related activities

Industry name	SIC 92 definition and classification
List A	Activities

Section F Construction	45.1 Site preparation
	    45.11 Demolition and wrecking of buildings; earth moving
	    45.12 Test drilling and boring
	45.2 Building of complete constructions or parts thereof: civil engineering
	    45.21 General construction of buildings and civil engineering works
	    45.22 Erection of roof covering and frames
	    45.23 Construction of highways, roads, airfields and sport facilities
	    45.24 Construction of water projects
	    45.25 Other construction work involving special trades
	45.3 Building installation
	    45.31 Installation of electrical wiring and fittings
	    45.32 Insulation work activities
	    45.33 Plumbing
	    45.34 Other building installation
	45.4 Building completion
	    45.41 Plastering
	    45.42 Joinery installation
	    45.43 Floor and wall covering
	    45.44 Painting and glazing
	    45.45 Other building completion
	45.5 Renting of construction or demolition equipment with operator
	    45.50 Renting of construction or demolition equipment with operator

List B	
Section C Mining and quarrying Subsection CB Mining and quarrying except energy producing materials	14.11 Quarrying of stone for construction 14.12 Quarrying of lime stone, gypsum and chalk 14.13 Quarrying of slate. 14.21 Operation of gravel and sand pits 14.22 Mining of clays and kaolin

Appendix Table 1A   Continued

Industry name	SIC 92 definition and classification
List B	Activities
Section D Manufacturing Subsection DB Manufacture of textile and textile products	17.51 Manufacture of carpets and rugs
Section D Manufacturing Subsection DD Manufacture of wood and wood products	20.30 Manufacture of builders' carpentry and joinery
Section D Manufacturing Subsection DH Manufacture of rubber and plastic products	25 Manufacture of rubber and plastic products 25.23/2 Manufacture of other builders' ware of plastic

Section D
Manufacturing
Subsection DI
Manufacture of
other non-metallic
mineral products
	26.11 Manufacture of flat glass
	26.12 Shaping and processing of flat glass
	26.22 Manufacture of ceramic sanitary fixtures
	26.30 Manufacture of tiles and flags
	26.40 Manufacture of bricks, tiles and construction
	      products, baked in clay
	26.51 Manufacture of cement
	26.52 Manufacture of lime
	26.53 Manufacture of plaster
	26.61 Manufacture of concrete products for
	      construction purposes
	26.62 Manufacture of plaster products for construction
	      purposes
	26.63 Manufacture of ready-mixed concrete
	26.64 Manufacture of mortars
	26.65 Manufacture of fibre cement
	26.70 Cutting, shaping and finishing in stone

Section D
Manufacturing
	28.11 Manufacture of metal structures and parts of
	      structures
	28.12 Manufacture of builders' carpentry and joinery
	      of metal

Appendix Table 1A   Continued

Industry name	SIC 92 definition and classification
List B	Activities

Subsection DJ Manufacture of basic metal products	28.22 Manufacture of central heating radiators and boilers 28.63 Manufacture of locks and hinges metals and fabricated
Section D Manufacturing Subsection DK Manufacture of equipment not elsewhere classified.	29.23 Manufacture of non-domestic cooling and ventilation equipment 29.52 Manufacture of machinery for mining, quarrying and constructionmachinery and 29.52/2 Manufacture of earth moving equipment 29.52/3 Manufacture of equipment for concrete crushing and screening and roadworks.
Section D Manufacturing Subsection DL	31.30 Manufacture of insulated wire and cable 31.50 Manufacture of lighting equipment and electric lamps
Section G Wholesale and retail trade	51.53 Wholesale of wood, construction materials and sanitary equipment 51.54 Wholesale of hardware, plumbing and heating equipment and supplies. 51.62 Wholesale of construction machinery
Section J Financial Intermediation	65.12/1 Banks 65.12/2 Building societies 65.21 Financial leasing 65.22/3 Activities of mortgage finance companies 65.23/5 Activities of venture and development capital companies 66.01 Life insurance 66.02 Pension funding
Section K Real estate, renting and business activities	70.11 Development and selling of real estate 70.12 Buying and selling of own real estate 70.20 Letting of own property 70.31 Real estate agencies 70.32 Management of real estate on a fee or contract basis 71.32 Renting of construction and civil engineering machinery and equipment

Appendix Table 1A   Continued

Industry name	SIC 92 definition and classification
List B	Activities
Section K Real estate, renting and business activities Subheading: Other business activities	74.20 Architecture and engineering activities and related technical consultancy 74.70 Industrial cleaning

2
The Capitalist Construction
Industry Labour Market

Introduction

While value added is the source of income and wages, labour is the source of value added. Without work, production could not take place. We now therefore turn our attention to the organisation of the construction sector and the labour processes involved.

The construction industry is a labour intensive industry, which has traditionally made extensive use of migrant labour, unskilled labour and the *informal* economy. There is also a high proportion of self-employed labour in the building industry. In this way firms in the construction industry minimise wage costs, provide minimal conditions of employment and in the process weaken trade union power. Moreover, much construction work, especially home improvement, is carried out by individuals on their own premises for their own use, on a 'do it yourself' basis.

One side effect of all these different methods of conducting construction work is the unreliability of official statistics related to construction. Government figures rely on registrations and reporting by firms and individuals and as such only cover a proportion of construction work carried out. The figures used in this book only need to be read as indicators of trends. The analysis of the data is based on *Housing and Construction Statistics* and the *United Kingdom National Accounts*, which make use of tables produced by the Department of Environment, Transport and the Regions, the Scottish Development Office, The Welsh Office, Department for Education and Employment, and the Office for National Statistics.

The trends in the labour force of the construction industry over the 1980s and 1990s relate to the patterns of employment and remuneration during these decades. Employment patterns over the 1980s and 1990s show a general deterioration in the terms of employment for a large proportion of the work force. Remuneration in construction was below the average of all other industries and services at the beginning of the 1980s. Improvements in pay relative to other industries and services were only temporary, caused by the growth in construction demand which climaxed at the end of that decade. Few statistics are kept concerning unemployed building workers. Unemployment can be viewed as a wasted resource, an aspect of under-utilisation, a broader concept, which can apply equally to materials, plant and buildings as well as labour. Labour unemployment refers to those people who are willing and able to take up paid work but are unable to find any. The measurement of unemployment is dependent on definitions used in data collection usually based on those who are registered as eligible for unemployment benefit. During the 1980s various changes were introduced which reduced the number of people entitled to register as unemployed.

In this chapter we examine the relationship between labour and capital in the production process. Capital is used to increase productivity over time enabling firms to increase profits, while remaining competitive.

The technological and social relations of production

Firms generate profits by paying less for all inputs (including labour) than the *gross* value of output. Alternatively we may say, that they generate profits by paying less for labour than the *net* value of output. In construction firms, wages and salaries form the major single component of their costs. Wages are determined by market forces in the labour market, while the value of net output is determined by the price and quantity of the goods or services produced and sold. The quantity produced in a given period of time is determined by labour productivity, which itself depends *inter alia* on the amount and type of plant and equipment used.

The relationship between people and technology is one of the most important themes in economics. The technology of production is constantly changing with the introduction of new methods, materials, plant and machinery. These technological developments bring about changes in the way people organise themselves in their economic relations in terms of methods of production and exchange, and social relations in terms of how individuals behave towards one another.

To understand economic and social relationships in the construction industry we must therefore look at the production process.

Chapter 1 dealt with the technological relationships between quantities of inputs and outputs of things. Labour inputs too were treated as if they were things. Labour, however, refers to people, not something inanimate. Technical relationships in production are between people and the things that they work on and with. These relationships have a qualitative aspect such as the knowledge of techniques carried by workers or by technicians and scientists, and a quantitative one concerned with the number of hours of labour, the number of materials and the number of machines.

The social relationships in production are between different individuals involved in the production process: workers, who hold technical knowledge, managers, and employers or owners. These social relationships may be said to be horizontal or vertical. Horizontal relationships are between the many different workers engaged in the joint production of an output. This is dealt with by economics under the heading of the division of labour. Vertical social relations arise between those people who do the work of production and those who employ them. This is dealt with under the heading of the employment relation and control within the labour process.

The social relations of production in any economy are structured by market relationships and employment relationships. Production is undertaken in firms. Within each firm the employment relationship rules. But production of a final product is also divided up between many firms. Between firms the market relationship rules. Thus, much of the horizontal division of labour takes the form of market relationships *between firms*, buying and selling each other's outputs to use as inputs. But within each of these firms we find vertical relationships implied by concentration and separation of control from the actual carrying out of the work.

Now, an employer and employee are in a sense engaged in a market relationship, in that one buys the labour power offered for sale by the other, but there is far more to the relationship than that. Crucial to it is that what is being sold is the right to command and direct the labour power of a worker, for a specified period of time. Moreover, what is bought is not an actual output or set of goods or service, but rather the potential to perform directed work that can produce a variable and open-ended output. It is thus quite unlike the usual relationship of exchange between the seller and buyer of a given set of goods, which does not involve a relationship of power and command.

In other hierarchical production systems, the power of command can come from monopoly of technical knowledge or from possession of state authority. But in capitalism it comes essentially from ownership and control of the physical inputs and durable means of production such as machinery and buildings.

The price of what is bought in the employment transaction is the wage per period of time. To the buyer the value or benefit of what is bought is open and depends on a whole variety of factors. First, how efficiently can that labour power be utilised within a technological system of production. The better the technology and its implementation, the higher will be the physical output per day of labour performed. Secondly, how intensively can that labour power be utilised. The harder or faster the worker works, and the less the unproductive time, the greater the physical output per day. Unproductive time is time that has been bought but in which no output-producing work is done. Third, the lower the price of non-labour inputs and the higher the price of outputs the higher will be the value added of any given amount of physical output.

Labour markets

In markets where goods are bought and sold, ownership is transferred from the seller to the purchaser. This is distinct from markets where goods are only hired. In hire markets, transactions are concerned with the use and services provided by owners. Ownership is not transferred. Construction industry labour markets are also only concerned with the hiring of services carried out by people on a variety of tasks, undertaken in the course of a construction project. This is a wage-in-return-for-effort bargaining process. Only in slave markets is ownership transferred. In building and civil engineering, there are labour markets for the services of different types and levels of skilled and unskilled workers in different locations. Through the labour market, labour is distributed to different construction firms and projects.

Conditions of employment and wage packages vary from market to market within construction. Labour markets can be found on street corners, on site or in newspaper and journal columns. They may be for casual unskilled labour, experienced craftsmen and women or for qualified architects, surveyors or engineers. This diversity of markets reflects the fragmented nature of the construction industry. Each market is distinct in its characteristics, and is a function of the supply of and demand for a particular kind of worker.

The supply and demand for labour

To avoid confusion it is possible to think of labour markets operating at four different levels within an economy. At a macro-economic level aggregate labour supply and demand deals with labour market issues affecting all labour and all industries. Second, at an industry level supply and demand for labour deals with labour as a resource shared by the firms within the industry and moving between firms. At a third level, there are also markets for specific occupations within a locality, in which firms with similar requirements compete with each other for labour with similar skills, knowledge and experience. Finally, labour markets exist for single employers offering specific types of work and individuals who are willing to offer job-specific skills, which may be of significantly less use to another employer.

The effective supply of labour at a certain wage is comprised of those workers who are willing and able to offer firms the current average (or even, marginal) productivity of existing workers employed at similar wages. As well as off-the-job training, this definition also needs to recognise learning curves for new employees and on-the-job informal training, especially important in construction. The effective supply of output is a function of the productivity of the effective supply of labour (a number of workers) at any given wage rate.

The supply of labour depends on the number of people with a particular type and level of skill, experience and training, who are willing and able to offer themselves for hire in return for a given contract of employment. At any one time, in most industries, most of the relevant supply will be offered by workers already employed by a firm and planning to stay there. Imbalances between demand and supply, however, have to be dealt with by hiring new employees or firing existing members of staff.

At the point of negotiating a contract of employment the employer is uncertain about the competence of the job applicant and the work effort a new employee may be prepared to make. Simply quantifying labour in terms of hours to be worked does not take into account effort, sometimes called *work intensity*. Work intensity can be viewed as the ratio of effort to wages. The greater the ratio of effort to wages the greater the intensity of work. The supply of labour, the quality of which is uncertain, is a quantity of hours people are willing to work. The number of hours worked is sometimes referred to as *work extensity*.

The number of people supplying the labour market in aggregate depends on demographic factors and the participation rate. The

participation rate is the percentage of the population offering themselves for paid work. The participation rate will therefore depend on
the age and sex distribution in the population, its state of health and
its social conventions. Whether a participant in the labour supply is
employed or unemployed in part depends on an individual's
confidence in finding work and their reservation wage.

The reservation wage is the minimum weekly rate of pay which is
regarded as just more attractive than continuing unemployment and
searching for a possibly better job. It compensates an individual for
giving up state benefits, if any, plus a margin to make the work effort
worthwhile. This is a subjective rate of pay which the individual
regards as a sufficient reward for giving up unpaid alternative activities,
including leisure. The reservation wage varies from person to person,
reflecting their domestic conditions and their assessment of the probability of finding a better job than any currently on offer. Individuals
know that the longer they are unemployed, the lower this probability
will become. Thus, a prolonged period of unemployment will reduce
someone's reservation wage. Social security benefits set a floor to it. In
general, as real wages rise, more currently unemployed people will be
attracted to apply for work as their individual reservation wages are
exceeded.

The demand for labour depends on a combination of several factors.
Firms require labour in response to demand for their output. In this
respect, demand for labour is seen as a *derived demand*. Labour is not
wanted for its own sake but only in response to demand for the products or services it ultimately provides. The larger the order book, the
greater the need for workers to carry it out.

In orthodox or neo-classical economics, however, the number of
people employed also depends on the technology used and the rate of
pay (including the indirect costs of employment). To explain the relationship between technology, pay and the demand for labour we need
to use the concept of capital intensity. Capital intensity is the ratio of
capital to labour and shows the volume of plant and machinery per
person employed. The greater the capital intensity of a production
process the less labour is required in order to produce a given output.
When wages are relatively high firms adopt methods of production
that are capital intensive, using plant and equipment to increase the
productivity of labour. An increase in work loads can be met through
the greater use of plant and machinery, rather than by employing
more people. When wages are low the incentive to substitute labour
with capital equipment is reduced and demand for labour is more

responsive to changes in the work load of firms. In a broad sense, there is something in this argument, though too much can be made of it, as we shall explain below.

A change in money wages in the construction sector will have two effects on demand for labour, one an *income effect*, the other a *price effect*, and these will work in opposite directions. Consider first the price effect. Since labour costs are the main part of production costs (especially when we consider that the costs of materials, etc. will themselves very quickly be affected by a change in wage costs), then a fall in money wages will, *ceteris paribus*, raise firms' profit margins. However, it is likely that lower costs will quickly lead to lower prices. Even if margins rise somewhat, it is unlikely to be by anything like the fall in costs. Now, lower prices for investment goods, such as the final output of the construction process, combined with steady or rising profit margins, ought to stimulate the total volume of demand for such goods. Thus total real demand for construction output rises and this requires an increase in construction employment. There is an increased derived demand for construction labour. When we look at the evidence, we do indeed find that when construction industry wages fall, this permits, and is accompanied by, a fall in tender prices. The weak link in the chain is in fact not there, but in the low current price elasticity of construction demand. That is, *unless other circumstances are favourable*, a change in construction tender prices (in either direction) may not be enough to persuade many potential investors in built environment projects to alter their planned volume of construction orders.

It is important to note that it is *relative* wages and prices (that is, construction wages and prices relative to other wages and prices in the economy) that will affect construction demand in volume terms. However, in an open economy with a large export sector, a general fall in all wages and prices across that economy (leaving relative prices within that economy unchanged) may stimulate demand in all sectors, via its effect on the volume of production in the export sector, and derived demand resulting from this.

If construction production as a whole is more labour-intensive than other sectors of final production then a general fall in wages relative to average prices (i.e. a fall in real wages across the economy) might stimulate construction demand, by causing a relative fall in construction prices compared to all prices.

Now, assume that in the short run construction labour productivity is given and unchanging. The labour input required to produce a unit of output is therefore constant, and the demand for construction

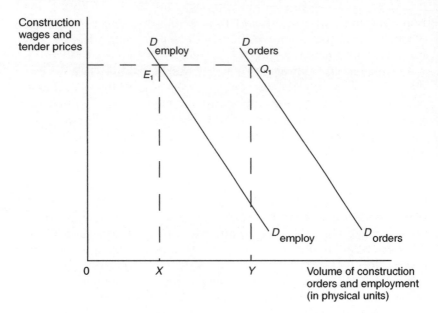

Figure 2.1 Derived demand for labour and the demand for construction (orders)

labour is simply a derived demand function of the demand for construction output. If the demand for construction output is given as Q_1 in Figure 2.1, and the demand for labour is E_1, then the ratio OX/OY is the inverse of average construction labour productivity. For example, if labour productivity is 40 physical units of gross construction output per worker per year, and the demand is for an annual output of 40 million of final construction output, then demand for construction labour will be 40m/40 = 1,000,000 jobs. Measuring in millions, Q_1 is 40 and E_1 is 1, the ratio of OX to OY is a constant 1 to 40.

The effect of wage rates on tender prices (and therefore to some extent, on construction demand) is the main price effect, in our view. We should note that neo-classical economists would look mainly for another kind of price effect. If wage rates change whilst the cost of owning and using a unit of fixed capital equipment stays unchanged, then the relative price of labour considered as one input compared to another rises, and a switching or substitution of production methods, to economise more on the now dearer input by substituting for it more of the now cheaper capital input would be predicted.

Budget lines represent the combination of purchases of labour and capital which can be purchased with a given sum of money. In

Figure 2.2, the budget lines Bp_0 and Bp_1 show the possible combination of capital and labour before and after an increase in real wages respectively. Each isoquant connects points representing different technical ways of combining inputs to produce the same quantity of output. The further an isoquant is from the origin the greater the output it represents. Hence, the point of tangency between a budget line and an isoquant represents the maximum output a combination of inputs on a budget line can achieve.

However, we would make two main points against this line of reasoning as applied to construction. First, the amount of fixed capital equipment used per worker in this industry (narrow sense of construction industry) is very low and shows no clear tendency to rise, despite long periods when construction wages have indeed been rising relative to the cost of fixed capital equipment. Second, we do not believe that production techniques are really as flexible or changeable in the short run (for example, a year) as this neo-classical analysis suggests they are. Firms are mostly pre-committed to using certain established production methods, and will only periodically, and in response to perceived

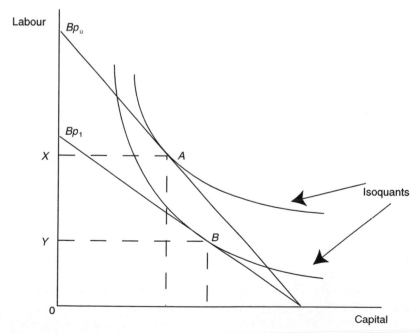

Figure 2.2   Substitution effect: labour and capital

long term trends, consider changing these. Thus the very conservatism of construction firms with regard to production methods works against the idea of capital–labour substitution (actually construction firms think of their key input–mix choice much more in terms of the on-site labour–prefabricated components mix, rather than the labour to machinery mix).

Now let us consider the *income* effect on construction demand of a general change in money wage rates. The basic Keynesian position is that the demand for consumption goods depends largely upon the level of wage incomes, whilst the demand for investment goods depends on the demand for consumption goods (this is called the *accelerator principle* and relies on the idea that, as for labour, a certain quantity of fixed capital input is required in order to produce a certain quantity of output). Consider the case of a fall in money wages. (The analysis of the effect of an increase in wages is the mirror image of this.) There will be a fall in consumption demand in money terms and, unless prices of consumption goods drop in line with wages (so that *real* wages stay the same – i.e. the purchasing power over goods of the wage paid is unchanged), therefore a fall in the quantity of such goods sold and, hence, produced. For construction demand, this would be felt directly mostly in a fall in households' demand for housing repair, maintenance and alteration. However, the accelerator effect means that demand for new construction capital goods would drop sharply, since industries making and selling consumer goods and services would have no reason to increase their capacity or stock of buildings faced with falling volume of production and sales. This would be the indirect income effect. We would expect the combined strength of the direct and indirect income-elasticity of demand effects to be quite strong.

The alert reader may at this point feel that they have spotted an incon-sistency in our analysis. In the preceding paragraph, we have assumed that the prices of consumption goods and services will not, in general, change in the same proportion as money wages have changed across the economy, including the consumer goods making industries. Whereas, when we were discussing the construction industry we stated explicitly that such a corresponding change is just what we think would (and does) happen. Prices in different parts of the economy do in fact behave quite differently – a point we shall develop in subsequent chapters, about *fixprice* versus *flexprice*. For the moment, we shall simply add that direct labour costs enter as a much more significant element in construc-tion firms' calculations of their full costs than is the case in many other (capital intensive, high overhead cost) industries.

Inefficient labour markets and non-clearing labour markets

The efficiency of a market is an idea developed by economists to describe the extent to which the same good or service always exchanges at the same price. Moreover for a market to be considered efficient certain other conditions need to be met. To the extent that a market is efficient this uniform price must approximate to the *market-clearing* price, which is the price that equates the quantities of supply and demand. The efficiency of a market is also determined insofar as this price equals the marginal cost of production. Finally, market efficiency depends on the price of each input or factor used being equal to its marginal opportunity cost, which is its efficient price in its next best alternative use. On this basis, out of all markets, it is probable that labour markets are amongst the least efficient, and that, out of all labour markets, construction labour markets are almost certainly not among the most efficient.

The inability of the labour market to operate efficiently in terms of neo-classical market economics is partly due to imperfect information on wages. Firms attempt to employ people on the basis of confidential wage bargaining. Only after a time lag do actual wages paid become common knowledge and anyone who is offered a rate of pay above an existing norm within a firm is discouraged from telling others because of the disruption to work it might cause. Firms do not openly advertise the actual wages paid. This lack of information on wage rates paid to existing staff or offered to potential employees by firms is the labour market equivalent of *price discrimination*. With the help of confidentiality between employers and employees and social barriers between employees, firms attempt to keep the wages of existing staff below the recruitment wage. This effect has been particularly important in professional services in construction, such as architectural firms, and also in the construction industry, where each gang on a site may be hired or paid at quite different rates.

There is an inertia in the labour market caused by the preference of people to remain with a current employer rather than to take on the risks of moving to a new firm and the costs and disruption of moving from one location to another. In order to attract labour from one firm to another, it is necessary to offer a higher wage in order to overcome the reluctance to move. Of course, this does not apply where employees are particularly dissatisfied by their conditions at work.

Nevertheless, because of the immobility of people and the fact that the labour market does not convey price signals in the manner described in neo-classical theory, the distribution of labour does not mirror the needs of employers or employees. Even at the equilibrium wage, where the number of people willing to work and the number of jobs available are equal, the number of people actually employed is less than this. This means that at the equilibrium wage, there are both unfilled vacancies and unemployed workers.

Negotiations over wages and conditions are rarely carried out by equal parties. When labour is in short supply, the labour market might be said to be a sellers' market and workers will be able to exact relatively high wages and advantageous working conditions. More usual in the labour market, however, is a buyers' market, in which employers can play one worker off against another especially if the labour required is unskilled and there is unemployment.

However, because of the relatively unskilled and unqualified nature of labour used in the construction industry (though not in other parts of the construction sector), and the ease of entry into its labour force, there are invariably downward pressures on wages from people who would otherwise be unemployed. Indeed one paradox of the labour market is that although the theory predicts that more people would offer their services at higher wages, in fact when unemployment is at its greatest and wages are driven down there is an increasing supply of construction workers, made redundant by recession in the rest of the economy and the number of people looking for construction jobs is at its greatest.

An important feature of the labour market concerns the time lags caused by the period of time required to train a workforce even after wages and therefore recruitment have risen. The period of training and therefore the responsiveness of the effective supply of labour to changing wages varies between the different labour markets. Time lags destabilise labour markets. In the short run wages rise because of the shortage of a particular skill. Higher wages in one period attract people into training courses in the hope of obtaining employment at the end. However, by the time training is complete the demand for labour has often changed. Even if demand remains high, the number of people entering the labour market is often in excess of demand for the new skills acquired, as individuals respond independently to the signals sent out by the high wages advertised in an earlier period. Consequently, far more people become qualified than are required and this surplus of labour causes wages offered by firms to new entrants to decline. This effect is obviously more important in the case of occupations for which the training period is lengthy, for example, architects,

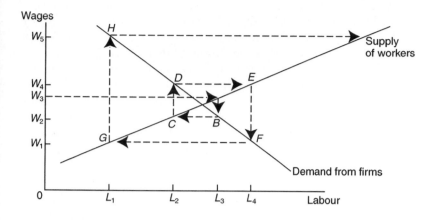

Figure 2.3   Diagram of the unstable cobweb effect

surveyors, engineers, and those manual crafts which still have apprenticeship-type training.

These effects can be seen in Figure 2.3, which demonstrates an unstable cobweb effect. Not all cob web effects lead to this type of disequilibrium. We simply use Figure 2.3 to illustrate the tendency for wages to continually rise and fall. Thus, if the wage level is at W_3, the number of people determined to work in this industry in the next period is L_3. This number drives wages down to point B because only at W_2 will employers take on L_3 workers. At this low wage rate only L_2 workers at C are willing to enter or stay in the industry, but employers are willing to pay W_4 wages at D as each firm attempts to ensure it has sufficient workers to complete its obligations to customers. As wages are driven up by competition between the employers looking for staff, more people enter the market attracted by the high wages on offer, but training causes a time lag until they are in a position to take up employment. Indeed at the new wage rate of W_4 so many people are prepared to work that their number at E exceeds demand from employers and a downward pressure on wages is repeated. As can be seen in the cobweb effect, therefore, wages do not necessarily tend to settle at the equilibrium wage.

In simple cobweb analysis, in each time period the supply curve can be thought of as a vertical line. Supply volume has already been determined by decisions based on past prices. Prices in this period will be set at the level where the demand curve meets the vertical line. Higher prices elicit no more supply in the short run, but *do* fundamentally affect the supply in the next period. In more realistic versions of the cobweb effect, there is some short run responsiveness of supply to price, but this is swamped by the stronger longer run response.

Taking a macro-economic view of the labour market, aggregate labour supply is relatively wage inelastic, in both the short and long term. In the short term a general wage rise will not be met by a greatly increased supply of labour. This is because of the time lag needed to alter attitudes to work and attract new categories of people into the labour market. Likewise, a general decline in wage rates may only serve to make people insecure and more determined to work and meet their financial obligations. They do not therefore withdraw from the market in significant numbers in response to a reduction in wages. That is not to say that unemployment does not change as a consequence. On the contrary, as noted above, both aggregate demand and total employ-ment would be reduced by a decline in wages.

However, at an industry level, there is variation between industries in terms of their relative wage elasticities of demand for labour. In some industries the labour to capital ratio and the labour to output ratio are relatively fixed in the short run by the technology. In any given industry the demand for labour depends on the demand for its output, and the ability to alter the technology and techniques used in production.

The dual labour market

The labour market in general has been described as a dual market. Two separate sets of conditions appear to divide the labour market into two distinct sectors. The first sector is called the primary labour market and is characterised by steady employment, often unionised, career struc-tures including training, promotion, relative job security, employment rights. In the primary sector remuneration is relatively high, and earn-ings are based on time, such as weekly or monthly rates of pay. The other sector is called the secondary labour market in which employees are employed on a casual basis, with few rights, no long term benefits, and usually paid on a piece rate or hourly basis.

Labour markets are therefore in fact concerned not only with rates of pay but also with conditions of employment. One of the weaknesses of traditional labour market analysis is that it emphasises pay and under-states the role of conditions or terms of employment. Conditions of employment can be seen as imposing costs on the employer which form a proportion of the total wage package. What matters in financial terms to the employer is not the amount received by the employee, but the total cost of employing an individual, which is the wage paid by the employer plus national insurance contributions, health and safety provisions, and the cost of other employee benefits.

The extra costs of employing people in the primary sector of the labour market, however, may be offset by the fact that employees are given an incentive to ensure the profitability of their employers. Given permanent employment, employees have a long term commitment to improving their performance through training and innovation, to ensure the competitiveness and survival of their firms. Their productivity is also improved through familiarity with their firm's methods, personnel and culture.

There is also a *cost of job loss* effect. The high cost of job loss for primary employees results from the fact they are paid a wage higher than they could readily obtain elsewhere, especially if job loss means they are forced to find alternative work in the secondary labour market.

In some labour markets, where labour mobility between employers within the primary market is very limited (for example, UK banks do not hire counter staff who have left jobs with one of the other UK banks), this effect is very strong, and creates real fear of losing one's job, which in turn employers use as the implied threat to extract extra compliance or effort from such employees.

However, in virtually all branches of the UK construction industry labour market, job turnover and labour mobility between employers are both high. The main component of cost of job loss then becomes the loss of wages during the period of unemployment likely to follow immediately on losing a job. This cost is further increased by employers' practices of offering lower recruitment wages to job applicants known to be currently unemployed.

The fixed costs associated with the employment of permanent staff impose a degree of inflexibility on firms who may not be able to shed sufficient labour when orders decline. Moreover, during recessions it may be possible to hire temporary labour at a cheaper rate than permanently employed staff, thus putting a firm with permanent staff at a disadvantage to its competitors.

Although casual employment in the secondary labour market appears mainly to benefit employers, workers themselves often enjoy the freedom to move to better paid work at short notice. To some extent they also enjoy the freedom to take breaks and holidays when they wish. Finally, many casual workers feel that they benefit from their treatment as self employed by the tax authorities. Broadly, during periods of labour shortage, workers may perceive the benefits of casual employment as greater than its disadvantages. This is reversed in periods of excess labour supply.

Undoubtedly many people in the construction industry enjoy being self employed. It is an industry which encourages the free-wheeling lifestyle of many of its workers and attracts some people for that reason. The casual work arrangement in construction is not a contract of employment but a contract for services and as such it tends to deny workers certain rights to compensation for injuries, and any rights to holiday pay, sick pay and employer's national insurance and pension contributions. Moreover, employers do not have the burden of administering employees' tax returns, while casual workers are left to carry out their clerical duties in their own time. Finally, the cost of periods of unemployment are borne by the state and casual workers rather than their employers, who would otherwise be liable for redundancy payments.

It is perhaps still too early to say what effect the 1996 Finance Act has had on construction firms' employment strategies. However, several reports have indicated that firms have begun to re-employ labour directly. The 1996 Finance Act has raised the threshold for self-employment and any firm employing a person later deemed not to have been self-employed, is liable for income tax payments in retrospect.

Construction labour

Statistics of labour in the construction industry are grouped into operatives, including skilled and unskilled labour, and non-manual staff working in contractors' or direct labour organisation offices, including administrative, professional, technical and clerical employees (APTCs). While APTCs are usually employed in the primary labour market, and operatives often work in the secondary labour market, there has been a trend in recent years to casualise APTC work in the construction industry by taking on staff only for the duration of a project. Indeed, similar and equivalent practices have begun to be implemented in other sectors of the economy as firms are *re-engineered* or re-organised so that fewer people are required to work on a permanent basis, on site, in offices or in factories and staff are encouraged to work on a freelance basis, on commission or on a piece rate. These trends, however, may only be a response of firms to recession and the opportunities provided by high unemployment.

According to Department of the Environment figures based on returns from contractors and public authorities, between 1981 and 1991 total construction industry manpower decreased from 1,606,000 people down to 1,526,000, a decline of 5 per cent. Closer analysis of the total number of people working in the construction industry reveals a significant shift to casual employment in construction. More specifically, construction employment statistics show that while the number of directly paid employees in the industry declined by 28 per cent from 1,218,000 people in 1981 to 878,000 in 1991, the number of self-employed workers increased by 67 per cent from 388,000 in 1981 to 648,000 in 1991. Self-employed workers are not entitled to minimum wages, holiday pay, redundancy money, sick leave, pension rights or any of the benefits accruing in theory if not always in practice to directly employed staff.

This rise in self-employment has probably had the effect of reducing annual productivity per worker due to the lower number of working days worked by the self-employed. Although the output of self-employed labour per day of work tends to be higher than direct labour, self-employed workers tend to take days off, or days working outside the industry, as well as needing time to search for work.

One reason for the low number of workers in trades unions on many building sites is the proportion of the labour force which is subcontracted. Labour only subcontracting is a form of employment commonly used in the construction industry. Firms can take on workers on a self employed basis. In fact, these nominally self-employed are little more than casual employees. In economic terms the difference between self employment and casual work is that in self-employment, the worker provides the materials and controls the equipment used and the methods adopted. Casual workers on the other hand have little input apart from their labour though they may own their hand tools.

Labour only subcontractors emerged out of the need of firms to reduce their direct employment to a minimum. When projects, or indeed tasks, come to an end, contractors have no legal obligation to continue employing labour only subcontractors, who are legally and for tax purposes self-employed. This means that firms do not have to pay for redundancy. When labour is not required to work on site there is no burden on the contractors to pay these workers. The labour only subcontractor must then find work elsewhere. In this way contractors pass the cost of sporadic working on to the economy as a whole.

Competition between contractors other than at times of boom or excess demand, means individual contractors cannot charge their clients by loading their tender bids to pay for unproductive time. As the tendering process is usually won on the basis of the lowest price, any firm with higher than average costs would become uncompetitive and eventually run the risk of going out of business. Minimising labour costs, to keep construction costs down to the level of competitors, helps to reduce the average tender bid price.

The paradox is that during periods of high demand for skilled labour, the wages of labour only subcontractors can rise beyond the employment costs of more permanently employed workers. The eventual cost of construction is then higher than it would have been had all labour in all contractors been employed directly by contractors in a conventional way. But these higher labour costs of using self-employed labour only subcontractors are more easily passed on to clients, because at such times competition is insufficient to prevent this. Since all contractors are faced with similar problems of finding labour, their competitive prices are then based on controlling their percentage mark up over these high costs.

In periods of recession the situation is reversed. Labour only subcontractors often then experience disproportionate cuts in their wage rates, as contractors find they can hire sufficient labour while offering reduced wages. In other words, labour only subcontracting allows contractors to be flexible and responsive to changes in demand, by helping firms to satisfy their short term requirements for labour.

Partly as a result of the prevalence of labour only subcontracting for manual workers, systematic training and apprenticeships have largely disappeared. The result is a skill shortage in the UK construction industry combined with a subjectively reported reduction in the quality of the workmanship available on site.

Health and safety considerations are often neglected and result in the industry having above average accident rates. It is a dangerous industry providing only a short working life for its labour force.

To Braverman (1974) labour only subcontracting systems suffer from problems of irregularity of production, slowness of manufacture, lack of uniformity and uncertainty of the quality of the product. But most of all, he wrote, the use of subcontracted labour inhibited change in the processes of production. In construction, new technology cannot be introduced if the labour needed to apply it has not been trained, is not known and is not hired on a permanent basis. Where subcontract firms specialise in new technological advances it is difficult to integrate change with traditional firms working on the same project.

However, in a period of full employment, firms would need to improve existing terms and conditions of employment as well as wages in order to keep and attract staff, and in order to be able to implement labour-saving new technologies. Otherwise they would face high labour turnover costs and be uninsulated from upward market pressures on wage rates.

The role of collective bargaining in construction

Industrial relations concern the processes of establishing working practices, negotiating pay and conditions, and resolving disputes between firms and their employees. In construction there are a number of approaches to industrial relations. One of these involves a formal set of arrangements known as the National Working Rules Agreement, (NWRA). This agreement between the trade union side and the employers' associations, relates to the pay and conditions of site labour and others on site, and is renegotiated and renewed annually.

However, in spite of these annual negotiations, the NWRA is not necessarily observed in practice. It is difficult to enforce, especially during recessions. For a variety of reasons, trade unions in the construction industry are weak. Site labour is difficult to organise. Individual workers move from site to site. The temporary and dispersed nature of sites means that unions cannot keep up with changes in the number, size and location of work places. The self employed nature of labour only subcontracting has also weakened the unions in construction. As a result trade union membership as a proportion of the total labour force is low. Moreover, there has been a protracted history of inter-union rivalry in construction as UCATT and the TGWU have competed with each other for members.

In construction, collective bargaining plays a relatively insignificant role. Although complicated annual negotiating machinery has been created, involving construction unions and employers' associations, the wages and conditions of employment are scarcely adhered to when local bargaining takes place. The wage rates agreed nationally appear to have little bearing on reality. Industrial relations in the construction sector can be viewed in terms of formal, informal and unformal agreements between firms and their workers (Druker and White, 1995). The formal set of arrangements are defined by the National Joint

Council of the Building Industry and by similar arrangements in civil engineering and some building trades such as heating and ventilating engineers. Separate arrangements are also discussed on a formal basis in Scotland and Northern Ireland.

Informal industrial relations occur in site level individual or group wage bargaining, where formal agreements are rarely used to determine wage rates or control conditions of employment. The actual level of wages and conditions of employment depend on local labour market conditions. If there is a shortage of labour, firms improve the wages and conditions offered, but respond to perceived surpluses of labour by reducing wages.

The third set of conditions found in construction concern arrangements totally outside the legal framework of employer and employee. This unformal labour sector concerns *labour only subcontractors*, who work on building sites as self employed individuals, as far as construction firms, the individuals and the tax authorities are concerned.

However, labour only subcontractors only enjoy self employed status in legal and fiscal terms. In fact, an economic analysis reveals that labour only subcontractors are really employees with no terms of employment. The distinction between employment and self employment has to do with the role of labour and capital. Capital includes machinery and materials. Self employed individuals can, in economic analysis, be independent producers *if* they own the means of production and control the materials used. In this case, they will be able to produce and sell an output, comprising goods or services, and will receive the residual after costs of production are deducted from revenue arising from sale of that output. For instance, a plumber who owns a van and equipment and provides the materials needed to carry out a job can be said to be self employed. Employees, on the other hand, only provide labour and use machinery owned by their employers to work on materials or components supplied to them by their employers. In this way employers control the amount of labour required, the terms of employment and manage the way work is carried out, determining the methods and activities of the workforce.

Within any work place there is a balance of power between the employer and the employee, depending on labour market conditions, which determine the ease of finding alternative work for the employee and alternative workers for the employer. The outcome of negotiations is a bargain concerning the wage/effort ratio. Employers and employees are drawn into a conflict over wages and work intensity.

Nevertheless, both the employers and long term employees have a common interest in the efficiency of the firm. Without efficient working methods, the firm would not survive to pay wages and generate profits. In any case, there are positive non-wage aspects of work for workers. These include job satisfaction, self-realisation, sociability and social definition, all of these being in some degree specific to the particular job context, and hence dependent on the continuation of their present employment.

In reality, in spite of the machinery of collective bargaining, wage negotiations are left to individuals to discuss with their employers. The actual wage-effort bargaining is done by workers with *subcontractors*. Main contractors have largely detached themselves from direct involvement in wage rates, productivity or unit labour cost. Instead, they use their market power, when they can, to push subcontractors' prices down. They then leave it up to subcontractors to pass on lower prices in the form of lower wages. In this way subcontractors may be able to maintain some level of profitability out of their operations. At other times, when market power shifts to the workers, subcontractors' increased wages and subcontract prices are not then absorbed by the main contractors but are passed on in the form of high construction prices to clients.

Construction workers work longer hours on average than their counterparts in all industries and services. In some years wages in construction can be below the average while in others the rates of pay can be higher, depending on demand for construction labour. Table 2.1 refers only to legal employees, excluding labour only subcontractors. The table compares manual workers in construction to all industries and services. While the weekly hours including overtime in construction between 1981 and 1997 were on average 44.9 hours, in all industries and services it was 44.6. Assuming a year to be 50 weeks, this difference translates into 14.5 hours a year more than their average counterparts in the economy, hardly a significant difference. However, this average figure does not take into account that in some weeks construction workers may be prevented from working due to weather conditions and unemployment, which implies that the variability of hours worked per week may well be greater in construction than in other more regular forms of employment. In the period 1981 to 1989, wage rates per week in construction were below average for all industries and services. After 1989, in five years out of the next eight construction weekly wages were above the all-industry average, but so were average hours worked per week.

Table 2.1   Full-time male manual workers' earnings and hours in
construction and in all industries and services (adult rates, not affected by
absence), 1981–97

| April each year | Ave. gross weekly earnings (£) | | Ave. weekly hrs incl. overtime hours | |
	Construction industry	All industries and services	Construction industry	All industries and services
1981	120.9	121.9	44.4	44.2
1982	131.4	133.8	44.6	44.3
1983	139.8	143.6	43.8	43.9
1984	149.4	152.7	44.3	44.3
1985	156.8	163.6	44.4	44.5
1986	167.2	174.4	44.4	44.5
1987	180.5	185.5	44.6	44.6
1988	195.8	200.6	45.4	45.0
1989	214.2	217.8	46.0	45.3
1990	245.7	237.2	46.0	45.2
1991	257.1	253.1	45.4	44.4
1992	258.9	268.3	43.8	44.5
1993	274.3	274.3	44.7	44.3
1994	277.4	280.7	45.1	44.7
1995	294.7	291.3	45.9	45.2
1996	308.2	301.3	45.8	44.8
1997	324.8	314.3	46.9	45.1

Source: Department of Employment, *New Earnings Survey* (annual).

Table 2.2 compares non-manual workers in construction to those in all
industries and services. The salaries paid to non-manual workers in
construction were consistently below those paid on average across the
economy. Indeed the difference in pay between construction and the
average was greater for non-manual than for manual workers. It is
important to note that this data relates only to non-manual (APTC)
employees of firms classed as construction industry contractors.
Unfortunately, we have no equivalent data from the *New Earnings
Survey* for employees of professional service firms in construction
(virtually all of whom will be APTC staff).

From Table 2.3 it can be seen that wage differentials between non-
manual and manual construction workers based on average gross
weekly earnings rose from 1.28 in the early 1980s to peak at 1.51 in
1992.

Table 2.2   Full-time male non-manual workers' earnings and hours in construction and in all industries and services (adult rates, not affected by absence)

April each year	Ave. gross weekly earnings (£)		Ave. weekly hrs incl. overtime hours	
	Construction industry	All industries and services	Construction industry	All industries and services
1981	152.0	163.1	39.5	38.4
1982	166.0	178.9	39.3	38.2
1983	184.2	194.9	39.7	38.4
1984	199.2	209.0	39.8	38.5
1985	208.4	225.0	39.8	38.6
1986	229.8	244.9	39.8	38.6
1987	243.9	265.9	39.8	38.7
1988	274.0	294.1	39.8	38.7
1989	312.6	323.6	40.3	38.8
1990	346.8	354.9	40.2	38.7
1991	368.2	375.7	40.0	38.7
1992	390.0	400.4	40.3	38.6
1993	401.0	418.2	40.0	38.6
1994	414.5	428.2	40.3	38.9
1995	431.6	443.3	40.9	39.0
1996	445.8	464.5	40.7	39.1
1997	460.0	483.5	41.3	39.1

Source: Department of Employment, *New Earnings Survey* (annual).

Over the period from 1981 to 1997, the difference in the hours worked by manual construction workers compared to manual workers in the rest of the economy was, as we have seen, only small. However, the difference between construction industry manual and non-manual workers' hours, was significant, equivalent to working 10 per cent longer per year.

Supply and demand in labour markets – a theory of construction wages and employment

We have seen that in any period the supply of labour of a particular type or skill is the number of people willing and able to offer themselves for employment at a given rate of pay and for a given set of conditions of employment. This definition does not consider the quality of workers, variations in their skills and knowledge or differences in effort each worker may put into the work. Nevertheless, a vague notion of

Table 2.3   Full-time male manual and non-manual workers' earnings and hours in construction (adult rates, not affected by absence), 1981–97

April each year	Average gross weekly earnings £			Average weekly hours including overtime hours	
	Manual	Non-manual	Wage differential	Manual	Non-manual
1981	120.9	152.0	1.28	44.4	39.5
1982	131.4	166.0	1.26	44.6	39.3
1983	139.8	184.2	1.32	43.8	39.7
1984	149.4	199.2	1.33	44.3	39.8
1985	156.8	208.4	1.33	44.4	39.8
1986	167.2	229.8	1.37	44.4	39.8
1987	180.5	243.9	1.35	44.6	39.8
1988	195.8	274.0	1.40	45.4	39.8
1989	214.2	312.6	1.46	46.0	40.3
1990	245.7	346.8	1.41	46.0	40.2
1991	257.1	368.2	1.43	45.4	40.0
1992	258.9	390.0	1.51	43.8	40.3
1993	274.3	401.0	1.46	44.7	40.0
1994	277.4	414.5	1.49	45.1	40.3
1995	294.7	431.6	1.46	45.9	40.9
1996	308.2	445.8	1.45	45.8	40.7
1997	324.8	460.0	1.41	46.9	41.3

Source: Department of Employment, *New Earnings Survey* (annual).

the supply of labour emerges in terms of a quantity which will vary depending on the wage offered. This applies in industry-wide labour markets, as well as to individual firms needing to employ additional staff and at the occupational level, in which more people with a given skill will be attracted into a particular line of work depending on the wages and conditions offered. Assuming all else remains unchanged, the higher the wage the greater the supply of labour.

The question is, how responsive is the supply of labour to a change in wages? The answer to this question depends on whether one is talking about the labour supply to an industry, a firm or a particular occupation. In general, the wage elasticity of labour supply depends on how quickly people respond to a change in wages. This depends on the flexibility of labour over time. Flexibility, in turn depends on the willingness of people to change jobs, their mobility, their level of skill and the length of training required to meet a given standard of work. It also depends on the number of people with relevant experience who are either employed in other occupations or currently unemployed or economically inactive. The supply of labour is therefore partly outside the control of firms, although they are in a position to influence the quantity of labour supply through expenditure on training.

On the other hand, demand for labour in most industries is largely dependent on the quantity of fixed capital stock, the level of technology embodied in the plant and machinery, and the level of effective demand. The quantity of fixed capital requires a certain size of workforce if the plant and machinery is to operate at its designed efficient rate of output. The quality of the plant and equipment actually used is determined by its age and the modernity of the technology. The more recent the technology used the less labour is required as new technology almost invariably substitutes machinery for labour. Thus the quantity of equipment and the quality of the technology used set the *upper limit* on the number of people firms would be willing to employ. The level of effective demand for the goods and services provided by firms sets the actual level of demand for labour as a percentage of the maximum possible at full capacity. In the long run, despite the labour saving bias of new technology and the substitution of machinery for labour, employment demand can be seen to grow with capital stock. Those economies with more rapid increases in capital stock tend to increase their demand for labour at a faster rate than economies with lower rates of fixed capital stock accumulation. The rates of growth of employment relative to the growth of capital stock, however, appear to depend on the extent of the labour saving bias in new technology. This model works well as an explanation of what determines employment levels in, say, the building materials industries.

However, it would be misleading to describe firms in the construction industry, especially builders, as holding large and expensive stocks of machinery and buildings. Nor are their profits highly sensitive to their ability to approach full capacity utilisation.

First, just as construction firms draw on a general pool of labour not permanently employed by them, they also draw on general stocks of plant and equipment not owned by them. The plant and equipment is in fact owned by firms in the plant hire industry. Thus, whilst the capacity of the building industry can be said to be determined jointly by the size of those industry wide stocks of labour and equipment, it is not meaningful to think of the maximum output of a construction firm as being in any sense determined by the size of the stock of capital equipment that it owns.

Secondly, whilst building firms do not have an *engineered-in* short run limit to their output and employment, this is much less true at the level of the construction project, the industry's nearest equivalent to a production establishment in other industries. Once a project's type, size and duration have been determined, then design becomes the main influence on the capital and labour inputs required. Moreover, the type and size of project is invariably determined by the client, outside the control of the construction industry. Where project specification is not in the hands of the construction firms (as is still mostly the case) then input requirements per project can be said to face the producer firms as given. It is then the number, size and the mix of project types that determines need for inputs, and hence demand for labour and plant.

In neo-classical economic theory, employment demand is more sensitive to the relative price of labour compared to the price of equipment than it is to effective demand for the final output. The neo-classical model considers the substitution of capital for labour as a set of alternative methods of producing a given output. Relative increases in wages in construction for instance, would lead firms to increase the mechanisation of the construction process to reduce their dependence on labour. Capital intensity in the production process would increase. Capital intensification has been achieved in building (as opposed to civil engineering works) only to a limited degree through improvements in the plant and machinery used on site, but more significantly by a shift towards prefabricated components, such as curtain walling and factory produced modularised systems. Thus the increased use of prefabrication techniques has been the response by construction firms to the increase in the wages of site labour relative to site productivity. Moreover, where mechanisation has occurred, its cost advantages over manual methods are usually far from marginal, and therefore unlikely to be removed by any feasible shift in relative input prices.

Civil engineering represents a different case where machine-for-labour substitution has been a more important process. During the labour shortages of the late 1980s some *leading* parts of the building industry experimented with capital-for-labour substitution, especially in materials handling. However, overall the construction industry statistics for the stock of plant and machinery show only relatively slow increases, possibly because construction plant leasing data, which in effect is the value of the supply of plant services without operators, is excluded from construction industry stock.

Formal and real subsumption of labour by capital

The purpose of discussing the role of labour in the construction process, in this context, is to understand the nature of the relationship between labour and employers. Construction can be seen as a labour process. It is a division of labour into specialised tasks, jobs and occupations, and a set of relationships between workers and managers, between those who execute tasks and those who define and control that work.

The separation of the conception of tasks from their execution occupies a central place in the literature of the labour process. Braverman (1974) and Bowles and Edwards (1993) provide the standard account of an historical process in which, at first, capitalists employ workers for wages. For a time, those workers continued to perform their work in the same way they did when they were independent self-employed artisans. In this situation, capitalists can be said to have formal authority but not effective power over the labour process. The workers still have some control over the way they work, what they do, how and at what pace.

Later, the real power of capital to re-design or re-constitute the labour process increased with the continuous introduction of new machinery and the division of the labour of design from that of execution. Moreover, the growth and concentration of scientific and technical knowledge about production methods in the hands of managers and technical staff further increased their control over the labour process.

The real subsumption of labour is reinforced by the inferior position of labour in relation to capital in the labour market. Because there is usually competition for work it is usually possible for employers to play one worker off against another. Even during periods of full employment when suitable workers may be difficult to find, alternative methods of working or imported substitutes are always available.

Capitalists are held to have an interest in re-designing labour processes in certain ways rather than in others. In particular, Braverman holds that they have a general interest in deskilling as much of the workforce as possible, in order to break craft control over working methods and practices. Deskilling work also maximises the potential supply of labour and thus minimises wage rates. Capitalists are also held to favour accentuation of the division of the workforce into an elite of technicians and designers, on the one hand, and a mass of unskilled or semi-skilled operatives on the other. This permits a combination of two distinct strategies to manage these two parts of the workforce.

The former group of elite technicians and designers are offered permanent positions with career paths within the firm, are given quasi-managerial status with salaries rather than wages, and various privileges including some autonomy from direct supervision. They are also relatively highly paid in return for an expectation of loyalty to the firm. They are expected to adopt the firm's goals as their own.

The latter group of operatives are instead managed through the rough and tumble of the labour market. They can be hired and fired and work under the constant threat of dismissal. Their wages are linked to individual output or their pace of work is controlled by direct supervision or the machine-pacing of work. In fact, international and inter-industrial comparative research has tended to reveal a more complex variety of strategies for the management of labour, on which, see Littler (1982) and, specifically on construction, Winch (1986).

The use of labour in the construction industry

Because of the temporary nature of employment in the construction industry, as work comes to an end on one project, labour is often laid off. From the point of view of labour, any improvements in efficiency and any increase in effort would therefore only hasten the prospect of unemployment, although some members of the workforce may be paid increased wages as a result until completion of their firm's subcontracted work package.

Subcontractors only stay long enough on site to carry out their own specialist functions. Building sites are in a continuous state of labour flux as subcontractors complete their work packages and others come on to site to start their tasks. Indeed the fluidity of the actual workforce on many building sites is even greater than simply the changing composition of firms operating there, since each subcontractor experiences its own level of labour turnover. People working for subcontractors enter and leave work on site, for other jobs for the same subcontractor, or to work for other subcontractors. Because of the constant churning of the labour force on construction sites, it is extremely difficult to control the actual production process on a day to day basis. Often site managers may not be able to give instructions directly to people working on site, without a subcontractor's permission. Problems can easily arise between one subcontractor and workers working for different subcontractors. Continuity is difficult to achieve as new people constantly arrive on site to carry out work.

The management of labour in general in modern industry is often described as Taylorist, Fordist, and post-Fordist. Taylor introduced the term *scientific management* to describe an analytical approach to work in which each task or element of a job a worker carries out is defined and then accurately timed to see if there are ways the tasks may be simplified. In this way the work can be deskilled, divided, mechanised and above all costed and controlled. The Taylorist approach tends to ignore the contribution of skill and the working conditions of the workers. The pace of work is determined by managerial analysis, breaking down work into its components, and recomposition of tasks by managerial planning and pre-calculation.

Then, as exemplified in the Ford car plants during the first half of the twentieth century, the pace of work came to be determined by technical factors, most usually the pace of the plant and machinery. Work was broken down into a series of sequential tasks and where possible a conveyor belt or assembly line type of approach was adopted. In factories people stood beside conveyor belts and worked on the product as the product moved by, each person carrying out repetitive simple tasks until eventually the mass-produced, standardised product was completed at the end of the assembly line. Fordism on construction sites has to be imagined as a reverse assembly line, in which workers move along a line of work stations, whilst the objects being worked upon remain fixed in space. Examples of machine pacing can be found in, for example, mechanical pouring of concrete.

By the end of the 1980s, manufacturing processes had been widely converted using computer aided design, management and manufacturing. The automation of the factory production process has led to an increased ability by manufacturers to respond flexibly to small batch production requirements. No longer do economies of scale depend on mass production techniques. Today manufactured products can frequently be efficiently customised at little or no extra cost of production. This post-Fordist form of manufacturing has meant that there is less need for people to carry out repetitive jobs on assembly lines for long durations. Instead robots and automated plant can be programmed to produce a variety of options quickly and cheaply. Labour is used to supervise plant and machinery rather than to assemble products, though of course many low paid, low skill functions remain.

Thus neither the Fordist nor post-Fordist 'image' of production seem to capture much of what has characterised construction sites as work places, or practices of construction management, in the twentieth century. Nevertheless, the basic idea is that a period of mass production, on the largest scale possible and often bringing together very large workforces under a single firm and at a single place, has given way to a period in which the tendency is towards greater flexibility, both in terms of adaptation of the product and in terms of organisation of the production process; and at this basic level the concepts of Fordist and post-Fordist production do seem to apply to production of the built environment, including the construction industry

Structure of the construction labour force

Until 1989 figures were kept by CITB relating to 15 categories of skilled operatives. These showed a marked decline in the number of skilled craft workers in construction throughout the 1980s. The one exception to the decline in numbers was the craft of electrician, which increased between 1981 and 1989 from 35,000 to 38,100 people. These figures are shown in Table 2.4. The data excludes public authorities direct labour and those employed by steel erection firms (although 'steel erectors and sheeters' is one of the categories), but includes those employed by firms manufacturing wooden industrial components.

The decline in supply of most types of skilled operatives relative to demand was sufficiently rapid and severe that serious skill shortages began to appear in the late 1980s and skill shortages have continued to be a feature of the construction industry. Several reasons for the decline have been suggested including a poor training system, demographic factors, the relative unattractiveness of construction work, poor career prospects and insecure work with periods of unemployment. Reasons for the decline in construction skilled labour demand have been the introduction of new technology and prefabrication.

Although Braverman's generalisations may be an oversimplification, the last few decades have witnessed the ending of the old 'craft' system in most of the UK construction industry in five respects. First, the apprenticeship training system has collapsed. Second, the ladder of opportunity from craft qualification to site manager no longer operates, since the influx of graduate trained APTCs. Third, there has been a growth in the ratio of APTCs to operatives, with the consequent increase in emphasis on management of the building process. Fourth, new components, tools and methods of construction have led to a rise

Table 2.4   Operatives (thousand) by craft, 1981 and 1989

	1981	1989
Carpenters and joiners	75.0	53.3
Bricklayers	33.2	20.2
Masons	2.3	1.7
Roof slaters and tilers	6.0	6.4
Floor, wall and ceiling tilers	5.2	3.6
Plasterers	8.7	5.2
Painters	36.0	23.9
Plumbers	21.7	16.6
Heating and ventilating engineering workers	13.4	11.8
Glaziers	2.4	2.1
Paviours	1.4	0.4
Steel erectors and sheeters	2.3	1.8
Electricians	35.5	38.1
Mechanical equipment operators	39.5	22.8
Other B and CE crafts and occupations	61.1	60.1

Source: *Housing and Construction Statistics*, 1981–1991, Table 2.2, from CITB data.

in 'fitting' or assembling tasks on site. Finally, there has been an increase in specialisation (and implied deskilling) even within occupations, for example 'first fix' carpenters.

Crafts or trades have tended to be transformed into occupations in which the majority of those working in that ocupation have not undergone (or have not completed) any formal process of qualification. In these new conditions, a construction worker (and thus the total workforce) can only effectively be identified *ex post*, as someone who actually does work in construction. Employers of operatives (mainly subcontractors) require of a new employee experience, perhaps, but rarely qualification. This in itself has greatly weakened the construction trade unions, through one or two craft-based unions (of whom the Electricians are probably the most important), continue to require workers on organised sites to be qualified, and exercise thereby some control over the relevant labour supply.

Concluding remarks

Thus, the overall package of working conditions and pay in the construction industry is somewhat below the average of all those working in all industries in the economy. This is not to say that construction workers are the lowest paid or that they work the longest hours on the worst terms and conditions. Indeed, for many women, an opportunity to work in construction would represent an improvement in their weekly wages, hours of work, and working conditions.

Within construction, as with the rest of the economy, the labour market is divided between those in the primary sector, who enjoy a package, consisting of direct employment, career structures, as well as holiday pay, sickness benefits and pension rights; and the secondary sector with casual, temporary and insecure employment, often associated with the duration of specific projects, without most if not all of the benefits given to those employed in the primary sector. When the projects, or the specialist contractor's work packages on a project come to an end, employment ceases.

However, labour relations consist of more than the contractual arrangements between firms and their employees. They are concerned with the way work is organised, with a view to increasing profits over time. To this end there is a continuous process of replacing labour with machines and prefabricated components, and deskilling the labour input at the same time. This not only reduces the number of people employed on site, but enables firms to lower wage rates below what they otherwise would be if skilled operatives were needed. For the individual the strategy needed to overcome the threat of low pay or redundancy, is retraining and the continuous acquisition of new skills and knowledge.

On the other hand, both ordinary and labour only subcontracting tend to reduce the real control of the main contractor over the content and detailed 'design' of the labour process. Firms cannot simply 'order' subcontractors to adopt new machinery or skills, or to perform tasks in new ways, in the same way that they could with direct employees.

3
Productivity and the Production of Profits

Introduction

In Chapter 2 labour relations in construction were discussed. Here we discuss the process of using employed labour to make profits. This is achieved through the productivity of labour and this will be considered at both the level of the firm and the industry. The productivity of labour will be seen to depend of the amount of plant and equipment used as well as the technology embodied in that equipment, but also upon a wide range of other variables.

Productivity

Productivity refers to the quantity of output per unit of labour in a given period of work. The output of a firm can be measured in terms of the physical units produced. However, this becomes rather meaningless in construction where each building or project is unique. Therefore the money value of output is used instead because it is necessary to find a common measure in order to find the total output in a given period of time or to make comparisons.

Labour productivity is then defined as the *value* of output per person per week or per year. However, prices can rise without any increase in production. Statistical tables of output frequently use current prices, which are based on the prices actually charged. These current prices therefore increase with inflation from year to year. For this reason, values of output at constant prices, which are based on the prices in one particular year, are a more accurate measure of annual changes in production. It is possible in principle to calculate real changes in construction output, using a construction price index to estimate the rise in construction prices from one period to the next. Different constant price deflators can be chosen either to eliminate the effect of general inflation occurring in the economy, or targeted more specifically to remove all price effects within construction so as to estimate changes in the physical volume of output. Eliminating the general effect of inflation is relevant for the study of firms' revenues, costs and profitability, since these are best measured relative to the rest of the economy. Removing construction price effects, taking account of specific price changes is relevant for the study of industry capacity utilisation and physical productivity.

Work on building sites is rarely steady throughout a project. Even on a day to day basis, labour productivity is reduced by the amount of time lost through waiting for instructions, breaks in work, discontinuities caused by late deliveries or preparation of work stations and so on. The proportion of the day that is wasted we call the *porosity* of the working day. The effort and motivation of the workforce and the effectiveness of management in organising work on site both have a significant influence on real productivity through their influence on porosity.

In measuring labour value productivity we are really interested in the average value added by workers to inputs such as raw materials and components. The value of sales is a measure of a firm's gross output. Gross output less the cost of all materials and building component and industrial services used is the net output or value added. Net output can be calculated for any firm in the construction sector involved at any stage in the building process:

$$GO - M = NO \qquad\qquad (3.1)$$
where

GO = gross output
M = cost of all materials and industrial services used
NO = net output.

On a building site, for instance, increasing the use of bought in off-site manufactured components would not necessarily change the gross output but it would cause the value added on site to decline. Prefabricated components bought in could increase the gross value of output per worker on site, while in fact the net value added by the on-site labour declined. This can be clearly analysed by reference to the following formulae.

Productivity can be measured using either gross output or net output. Productivity per person is found by dividing a firm's output per period of time by its labour-input in time units, measured as the number in its workforce multiplied by the labour-time per worker. If we assume a standard time for a working day, then:

$$GP = GO/L \qquad\qquad (3.2)$$
where
$$GP = \text{gross productivity per person per day}$$
$$L = \text{number of worker days}$$
or,
$$NP = NO/L \qquad\qquad (3.3)$$
where
$$NP = \text{net productivity per person per day.}$$

The distinction between gross output and net output in the building industry is important in understanding the contribution made by firms and labour on site to the value of the final building price. Productivity based on gross output per worker on site has increased greatly in recent years due to the increase in prefabrication and the consequent decline in the number of workers needed on site. At the same time, the value added to inputs by firms on site has declined as a proportion of gross output. Nevertheless, the net productivity of site labour has continued on balance to rise, partly because of new techniques of construction, improved site management, and partly because of the introduction of improved plant and equipment.

Figure 3.1 shows employment and output in the construction industry. *Gross* construction output is shown at constant 1995 prices. From Figure 3.1, it can be seen that gross output per manual worker rose from £34,100 in 1983 to £47,900 in 1997. The rising trend in gross productivity continued even after total construction output peaked in 1990 at £58 billion. The number of manual workers employed in the industry declined from 1.5 million people in 1990 to less than 1.2 million by 1993. Their gross productivity rose by 30 per cent between 1983 and 1993 or by over 40 per cent between 1983 and 1997. However, the increase in their productivity can in part be accounted for in terms of an increase in the use of prefabricated components and changes in the types of building being undertaken compared to work in the early 1980s.

Wages influence workers' effort or work intensity, and thus their productivity. But the intensity of work depends not only on the balance of supply and demand in the labour market, but also on several other factors, which influence matters inside the work place. For example, in

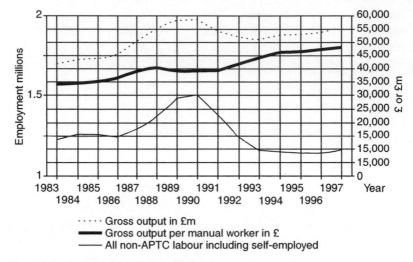

Figure 3.1 Manual labour and construction output, 1983–1997
Note: Output given at constant 1995 prices.
Sources: *Housing and Construction Statistics* (1984–1994, 1987–1997), Tables 1.6 and 2.1.

construction, effort is influenced by management methods such as the employment planning programme and by re-hiring practices, controlling workers' moves from one project to the next. Motivation is achieved by linking wages to effort in the short run, and is encouraged by the residual work ethic of autonomous craft labour of 'a fair day's work for a fair day's pay'.

Learning curves measure the rate at which output increases per period of time because people become more proficient with practice at the same set of tasks and problem solving. Continuity of work creates learning curves which improve the efficiency of labour. Unfortunately the one-off design of projects and the high labour turnover found in the construction industry limit the effect of learning curves. On building sites it is usual to find different specialist subcontractors in the same trade, with their own staff, working on site during different phases of construction. Their staff may be sent to work on other sites even during a contract or indeed may leave for alternative employment, so that not only do different firms work on site at different times but within these firms there is further staff turnover, as new workers are brought on site to replace those who have left. Moreover, efficiency gains from learning during a project can hardly ever be carried to the next project.

Average net productivity

Turning from the issues of labour productivity on individual sites to the issues of labour productivity in the construction industry as a whole, it is still the productivity of individual workers which contributes to the output of the industry. Thus the output of the industry is equal to the number of people in the workforce multiplied by their average productivity, as follows:

$$Q_I = aL_I \qquad (3.4)$$

where

Q_I = industry net output
a = net productivity per worker
and $\quad L_I$ = industry workforce.

$$a = f(b,(K/L),e_n,i) \qquad (3.5)$$

where

b = technical knowledge embodied in equipment
K/L = capital stock per worker
e_n = technical knowledge not embodied in equipment
and
i = work intensity.

From official statistics, it is possible to estimate both the value of net output of an industry and the number of workers it employs. These figures can then be used to yield a crude measure of increase in productivity on average for the industry as a whole.

However, a more refined estimate of productivity would also take into account the number of hours worked to achieve a given annual output. Just such an estimate was carried out in a report by NEDO (1986). Productivity in the building industry was calculated by dividing construction output by the number of people employed in the industry multiplied by the average hours worked. Table 3.1 reproduces the relevant finding of that study, in which it can be seen that net productivity increased by only 1.2 per cent from 1975 to 1980.

As in other sectors of the economy, productivity (measured as output per worker per year) in the construction industry has been rising, though not steadily nor as rapidly. Between 1975 and 1989, the rate of growth of productivity in construction was approximately 1.4 per cent per annum, although between 1978 and 1983 productivity actually fell, because output declined by about 5 per cent while the workforce

Table 3.1   Productivity index, 1975 and 1980

	1975	1980
Construction output (1975 = 100)	100	96
Employment (million)	1.27	1.23
Average weekly hours worked per person	45.2	44.0
Productivity index[1]	100	101.2

[1]Productivity index = O/(E × W) where *P* = productivity index
O = total construction industry output (1975 = 100)
E = employment (millions)
W = Average weekly hours worked per person.
Source: *Construction Industry into 90s* (London: HMSO, 1986).

rose by just under 5 per cent. The main rise in productivity occurred in the period between 1984 and 1989, a period of expansion of total output.

The productivity increases in the late 1980s may have been partly due to changes in the type of output, as productivity in the construction sector is primarily affected by the type of structure and the complexity of the design. The rise in the number and value of private sector office developments, especially open plan offices, raised productivity partly because of the high degree of repetition and the simplicity of core and shell construction. Construction work in public sector projects tends to be less productive than private sector developments, building less than civil engineering, and repair and maintenance less than new construction. The increase in productivity does not necessarily imply improvements in the techniques of production for similar projects.

While the highest productivity in the construction industry is to be found in civil engineering, the lowest productivity occurs in those firms undertaking repair and maintenance. The low productivity found in repair and maintenance work is partly due to the high porosity of the working day, which is caused partly by the amount of unavoidable non-productive time spent waiting, the impossibility of planning tasks efficiently, lack of repetition and much moving between projects.

Alternative estimates of level and rate of change of construction productivity

Several methods are available for calculating productivity. Each method comes from a different source of data. Method 1 uses value added as found in the *Blue Book* divided by the *Labour Force Survey* workforce, the sources of data being the ONS. Method 2 also adopts a value added approach, but uses *Census of Production* data. In this method COP value added is divided by COP total employment. Three different measures of productivity can be found using DETR sources. The first (method 3A) uses output data from HCS Chapters 1 and 2. This output data is divided by manpower (including both private and public sectors), also from HCS. Alternatively, method 3B uses the same output and manpower data but only that pertaining to the private sector. The third method (method 3C) is taken from HCS Chapter 3, which contains the Private Contractors' Census. This census provides the value of work done and total employment by private contractors and again the former is divided by the latter.

The latest published COP data at the time of writing was for 1995. It is therefore for that year that we make the comparisons. All output data used in Table 3.2 is at 1995 prices.

Thus we have estimates of productivity, all from government statistical sources for the same year, of which the highest is no less than three times the magnitude of the lowest. The best concept with which to measure the 'output' numerator in a productivity ratio *for purposes of comparison with productivity in other industries* is undoubtedly *value added*. Of the two estimates of value added (*Blue Book* and COP) there are clear grounds for preferring the *Blue Book* figure. The COP figure is not subjected to the cross-checking for consistency with the whole set of national accounts undergone for all figures published in the *Blue Book*. Moreover, by its nature, the COP figure is more likely to underestimate as a result of incomplete coverage – and indeed we see that the COP estimate for value added is somewhat smaller. It is possible that the *Blue Book* seriously understates construction value added – but we are not aware of any conclusive work showing that it does. The onus would seem to us to lie on those who would propose the use of any of the other methods (especially, those based on DETR output estimates) first to demonstrate, by research, that the *Blue Book* is inaccurate.

As for choice of the 'labour input' denominator, we consider the most appropriate choice to be the *Labour Force Survey* (LFS) – mainly because it is consistent and comparable with figures from the same source for other industries. Thus, we recommend, as the statistical consumer's 'best buy' for the purpose of comparing construction productivity to other industries, the estimate produced by Method 1 – which happens to give the lowest estimate.

Table 3.2   Level of productivity, alternative estimates, 1995

Method	Output	'Employment'	'Productivity'
	£ million	000	£000
1	32,948[1]	1,783[2]	18.48
2	22,949[3]	968[4]	23.71
3A	52,643[5]	1,375[6]	38.29
3B	48,942[7]	1,250[8]	39.15
3C	40,724[9]	751[10]	54.23

Sources:
ONS (1998) *UK National Accounts, 1998 edn.*
DETR (1998) *Housing and Construction Statistics, 1987–97.*
ONS (1998) *Business Monitor PA 1002: Production and Construction Inquiries, Summary Volume 1995.*
ONS (1997) *Labour Force Survey: Historical Supplement 1997.*
[1] From *Blue Book*, Table 2.3, 'Gross value added at current basic prices: by industry'.
[2] Derived by averaging total employment in construction reported in each of 5 quarterly surveys covering part of 1995; seasonal differences are very small.
[3] 'Gross value added at factor cost'.
[4] Total employment = 'APTC and operative employees on payroll' plus working proprietors of firms on Inquiry register.
[5] From HCS, Table 1.6, 'Output: output by contractors including estimates of unrecorded output by small firms and self-employed workers, and output of public sector DLOs – 1992 SIC'.
[6] From HCS, Table 2.1, 'All manpower'.
[7] From HCS, Table 1.7, 'Contractors' output, including estimates of unrecorded output by small firms and self-employed workers – 1992 SIC'.
[8] From HCS, Table 2.1, 'All manpower *less* public authorities' employees in employment'.
[9] From HCS, Table 3.3, 'Private contractors: work done, 3rd quarter', multiplied by four to convert to annual equivalent.
[10] From HCS, Table 3.4, 'Private contractors: total employment (including working proprietors)'.

The other main purpose to which productivity figures are put is the measurement of comparative rates of productivity growth. This is discussed in the next section.

Rate of increase in productivity: alternative estimates

Table 3.3 shows *Blue Book* value added/LFS employment using method 1 based on the ONS data. A second method, based on method 3A (p. 67) using the DETR data from HCS Chapters 1 and 2 is illustrated in Table 3.4 showing output at constant 1995 prices/manpower.

Thus, we have alternative estimates showing productivity increased by 15 per cent or 25 per cent over an 8 year period. Additionally, one

Table 3.3 Rate of increase in productivity: method 1, 1989–97

Year	Value added index (1995 = 100)	Employment number	Employment index	Productivity index	% change on previous year
1989	108.2	2137	122.9	88.0	
1990	111.3	2051	117.9	94.4	+7.2
1991	102.3	1879	108.1	94.6	+0.2
1992	98.3	1738	99.9	98.4	+4.0
1993	97.1	1572	90.4	107.4	+9.1
1994	100.8	1779	102.3	98.5	−8.3
1995	100.0	1739	100.0	100.0	+1.5
1996	101.5	1735	99.8	101.7	+1.7
1997	103.8	1777	102.2	101.6	−0.1
1989–1997					+15.5

Table 3.4 Rate of increase of productivity: method 3A, 1989–97

Year	Output index (1995 = 100)	Employment number	Employment index	Productivity index	% change on previous year
1989	110.4	1806	131.3	84.1	−1.0
1990	110.9	1832	133.2	83.3	−0.1
1991	102.8	1698	123.5	83.2	+7.5
1992	98.8	1520	110.5	89.4	+5.6
1993	96.8	1410	102.5	94.4	+5.3
1994	100.1	1384	100.7	99.4	+0.6
1995	100.0	1375	100.0	100.0	+2.7
1996	102.3	1370	99.6	102.7	+1.9
1997	105.4	1384	100.7	104.7	+24.5
1989–97					

might construct various composite indices, at the preference of the analyst – for instance, one combining ONS value added series and DETR manpower series. This would show value added falling by 4 per cent 1989–97 and employment falling by 23 per cent, and a productivity increase of 25 per cent. One should note that, for the period 1989–97 at least, the two methods used above of measuring output, though they show very different absolute levels, show broadly comparable rates and year-on-year pattern of output change. As between the numerator and the denominator in these productivity indices, the greater difference concerns the changes in the size of the construction work force.

Finally, by selecting an appropriate price index and using it to deflate series for output at current prices, one could also construct productivity indices based on PCC or COP output series. Price indices, however, are rarely completely accurate themselves, and moreover one faces another choice between different price indices, which in itself will yield different resulting measures of productivity change. The constant-price output series published by ONS and DETR are, of course, themselves derived using price indices – but at least in this case these are indices devised for this purpose, and therefore perhaps a little more reliable.

The main point we wish to emphasise is that there is no 'one best answer' to the problem of measurement of construction industry output, employment or productivity – but beware arguments which *assume* that the rates of change or levels of these variables are simple, well-known facts.

Marginal productivity theory

Marginal product may be defined as an increase in total output due to employing an additional unit of labour, such as an extra person or even an additional hour of overtime. Assuming that everything else, such as plant and machinery, remains the same, the law of diminishing marginal productivity states that, if labour is increased in order to increase output, eventually the increase in output, per extra unit of labour employed, will begin to decline.

According to the theory, diminishing marginal productivity occurs because the quantity of capital is held constant. Bottlenecks arise as the designed capacity of equipment is approached. Once machinery is operating at full capacity, output cannot be raised no matter how many people are employed. The construction site equivalent of this occurs when the number of workers is increased relative to the size of a site, and therefore its capacity to supply work *places*.

The traditional neo-classical argument derived from the law of diminishing marginal productivity is that, for profit maximisation, the wages paid to the last worker employed are equal to the marginal product. Moreover the theory assumes that the wages paid to the last worker are the same as the wages paid to other workers. The same is true in neo-classical theory for all factors of production, land and capital. For profit maximisation, the marginal product of the last unit of a factor of production must be equal to its unit cost. Otherwise, the firm could improve its performance by shifting resources in favour of a factor whose ratio of marginal product to unit factor cost is the greatest.

One practical application of the law of diminishing marginal productivity concerns the employment of additional labour. The question to consider is, what difference to total output would occur if one more person were to be employed? If the value of total output were to rise more than the cost of extra wages, then it would be economically viable to employ the extra labour. However if the same expenditure could purchase machinery which would raise the value of output by an even greater amount, then resources would be shifted towards the purchase of equipment.

It is often argued that the marginal productivity of labour is the determinant of the level of employment and the wage rate, since a profit maximising firm can afford to pay the last worker (and hence all other workers employed in the same capacity) an amount equal to the marginal product. For instance, if the value of the marginal product is equal to, say £200, then an employer could pay £200 to the last worker to join the team. If wages were to rise, so the argument goes, then the value of the marginal product would have to rise and so workers would be laid off until the value of the marginal product were once again equal to the new higher wage.

However, Robinson and Eatwell (1973) point out that the marginal productivity of labour does not explain the level of the wage rate. Thurow (1990) raises several objections which are of particular relevance to the construction labour process. Apart from the practical difficulties of calculating marginal productivity, it is not clear, for instance, if groups of workers as a whole are paid their marginal product or if it is paid to individual workers in return for their marginal product. How, for instance, would the marginal product of a gang of bricklayers on a building site be determined? Different members of the gang, such as the hod carrier, perform different tasks. Without their individual contributions even the project as a whole would be incomplete and valueless. Indeed in practice, to value the contribution of the bricklaying gang, the work is put out to tender as a work package. The job is then let to the gang offering the lowest bid. Their eventual wage payment is based not on their marginal product but on competition between workers in the labour market. Similarly, wage differentials cannot be explained by marginal productivity theory, which assumes that all workers in a group receive the same wage.

The production function

The quantity produced depends on the amount of labour and capital employed as well as the quantity of materials used.

$$Q = f(L,M,K) \tag{3.6}$$

where

$$Q = \text{output in a given period}$$
$$L = \text{labour employed}$$
$$M = \text{materials}$$
$$K = \text{capital employed.}$$

Most important are the ratios of output to labour as a measure of productivity and capital to labour, which can be used to compare the capital intensities of various industries. In this way productivity can be seen as a function of capital intensity. Thus,

$$P = Q/L \tag{3.7}$$

where

$$P = \text{productivity}$$

and

$$Q/L = f(K/L)$$

It would seem obvious that the value of capital stock owned (in a firm or in an industry) can be calculated by using the historic cost of purchasing the stock of plant and equipment less depreciation. However, the cost of new machinery can drop. This could make it possible to increase the amount of machinery without raising the value of capital, wrongly implying that there would be no improvement in productivity. In fact, it would be possible for workers working with cheaper equipment to produce more than workers working with expensive older plant. This is especially the case with computerised equipment.

Nevertheless, provided that the age, price and quality of equipment is taken into account, the value of capital stock owned per employee can be useful for the purpose of making inter-company comparisons.

The production of profit

The purpose of the production process in capitalist firms is to create an output and sell it at a profit. Managers continually seek new or redesigned ways of organising production in order to increase the profits of their firms. To understand this process overall, we need to define a set of terms and apply them to a firm. This model of production involves investment, output, wages and gross profit. *Investment* refers to the capital required for materials, plant and equipment necessary for the smooth running of a firm, planning to produce a certain target level of output. *Output* is what is actually produced in any given time period, measured in monetary terms, and may differ from planned output. *Wages and salaries* are paid to all employees, who spend their time in the service of the firm and receive a payment in return. The excess of output over the payment for intermediate inputs (including materials and manufactured components and equipment), and over wages and salaries and indirect taxes provides the gross trading *surplus* or gross operating profit, from which profits are taken. Thus

$$S = R - (W + M) \tag{3.8}$$

where

$$S = \text{surplus}$$
$$R = \text{revenues from sales}$$
$$W = \text{wages}$$
and
$$M = \text{cost of purchases, materials, etc.}$$

Profit does not appear in this equation because profit is part of the gross surplus derived from the production process. To arrive at the amount of net profit we must deduct capital consumption or fixed capital costs, interest payments, rents and administrative overhead costs. To then get the rate of profit on capital *owned* we also need to look at the ratio between this net profit and capital owned, which is itself a complex variable in relation to capital *used*. Wages are determined in one way through the labour market, while the value of labour productivity is determined in another way, through the market for the output, the technology used and so on. Amongst other things, profit depends on driving wages and average value productivity apart.

Total investment in a firm extends beyond buildings and plant and machinery. It includes working capital, which itself includes net current assets, and cash or current bank accounts for the day to day functioning of a firm. This cash enables the firm to have sufficient funds to meet its wages and short term liabilities, if not matched by other current assets, such as the firm's debtors. The minimum working capital requirement can be defined as the minimum net current assets necessary to enable production to take place. Physical assets such as stocks of finished goods, and financial assets such as claims on clients for payments due are called *current* assets because of the reasonable assumption that they will be turned into cash revenues within a short period of time.

Unit labour costs and labour and profit shares in output

We are now in a position to build our model of the production process. Labour costs per unit of output produced are affected by a combination of labour productivity and wages. Let us begin by assuming that extra output can be sold without reducing price. If productivity and wages increase at the same rate, then labour costs per unit of production would remain the same. A 10 per cent increase in output per hour would cover a 10 per cent pay rise, without any increase in the labour cost per unit of output produced. Only when wages rise faster than increases in productivity do unit labour costs rise. Naturally, it follows that if labour productivity rises faster than wages, the result is a drop in unit labour costs.

As labour is a high proportion of the costs of production, any reduction in the cost of labour per unit of output will tend to reduce costs significantly and raise gross trading surplus. If productivity has been increased by investing in more equipment, then the question for the firm is whether this increase in gross trading profit is also enough to increase the return on capital employed. The process of increasing the ratio of capital to labour is known as capital deepening, which increases the capital intensity of a production method.

Capital intensity is not the only measure of the advancement of an industry, because the vintage of technology used and the rate of technical change are also important. Because of the expense of machinery, high capital to labour ratios are often associated with predictable demand and monopoly control, since high profit margins and minimum risk are needed to undertake, finance and sustain investment in plant and equipment. Capital intensity is high in utilities such as gas, water and electricity supply and in the chemical industry, but relatively low in the clothing industry and construction.

However, it is not only capital intensity which raises productivity but also the technology embodied in the equipment. New plant and machinery will tend to be more productive than older *vintages* because of continual research and development. New plant and equipment incorporates the latest technological developments and innovations.

Moreover, it is very problematical to measure the value of the *use* of machinery or capital equipment by a firm or on a project. This is a very good reason for not trying to include a figure for capital consumption in the unit cost of production.

Division of labour in the construction industry

Specialisation of work has come about as a result of the economic benefits of what Adam Smith called 'the division of labour'. The division of labour describes the segmentation of production into separate tasks. By working as a team and each one specialising in one stage of a production process, a team of workers can increase its total output dramatically over that achievable by the same number working independently with each person therefore having to perform all tasks.

First, the amount of unproductive time – time lost in moving from one task to another, setting-up and closing-down – is thereby reduced. Second, through repetition and learning, each specialist worker becomes more adept and expert at their task – finds the 'one best way' of doing it, and perhaps discovers new ways. Third, specialisation permits (makes economical) the making of 'dedicated' investments. That is, investments whose only benefit is to improve efficiency of performance of one specific task. These can either be investments in the ordinary sense of the word, i.e. purchase of machinery, or investments in specialist training. Dedicated investment may involve indivisibilities. In this case, its economical use also requires a certain scale. By specialising, one work establishment can more easily achieve the volume of work in its specialism required to utilise fully an indivisible and dedicated machine or other investment.

But a complex division of labour does not only affect efficiency. The point is that the division of labour enables an employer to enter the production process as an organiser and controller of the workforce. By co-ordinating the volumes of output of workers engaged on each task to the requisite proportions, such employers can also minimise the need to hold idle stocks of unfinished goods or work in progress.

There is indeed an argument that the power of capitalists over the production process has been boosted by periodic drives to reduce the cost of holding stock, as much as by drives to raise productivity. An early example was when largely 'wet' methods of construction (which required part finished buildings to be left to dry) were replaced by 'dry' construction, even though this meant replacing cheaper inputs by dearer ones. For capitalism, time is money.

Even in 1832 Charles Babbage had pointed out in his book, *On the Economy of Machinery and Manufactures*, that by dividing detailed work between individuals, employers could reduce the level of skill needed and the amount of training given to workers. In this way employers were able to lower the wage rates they offered. Clearly the interests of the workforce are in conflict here with those of their employers, in that *excessive* division of labour reduces job satisfaction and increases worker alienation, as well as reducing the chance of autonomous work.

In construction, specialisation has led to the creation of a workforce with many crafts and professions. Indeed each craft and profession is itself further subdivided into even more specific tasks or areas of work. For instance, carpenters may specialise in hanging doors, inserting windows or fitting kitchens, bathrooms and bedrooms. Surveyors may specialise in the valuation of property, estimating quantities for builders or advising clients on building cost.

Specialisation in a single, narrow set of tasks by (the workforce of) enterprises in construction is carried to its highest level by subcontractors, and by some manufacturers of components. If the advantages of specialisation were the dominant force, then the sector would consist chiefly of firms each specialising in one part of one *trade* or one step in a manufacturing process. In fact, as we shall see, there are at least two other forces at work shaping the range or scope of firms' activities. One is the drive to reduce risk by spreading a firm's activities between several markets, leading either to a kind of 'generalism' (as, say, in main contractors who tender for almost any kind of construction regardless of the technology and tasks involved) or to diversification. The other is the drive to achieve competitive advantage or monopoly power over rivals, or reduce the costs of inter-firm transactions, by vertical integration. Thus road building contractors find it advantageous also to control local production of aggregates and roadstone; or estate developers also to act as house builders.

Subcontracting may increase the net productivity of those employed by main contractors, if subcontractors are used to carry out the low productivity jobs. The overall productivity of all the firms on site may nevertheless remain unchanged. By subcontracting the low productivity jobs, main contractors can take an increased share of the value of a contract per employee.

Concluding remarks

Firms produce a gross trading surplus by selling the output of the firm at a price greater than the amount paid for material inputs and labour. This surplus is needed to pay for capital consumption, administrative overheads, rent and interest. The size of this surplus depends on the productivity of labour and this in turn depends on the degree of capital depth of the production process, and the vintage of the technology used. It also depends to a large extent on how work is organised and managed, especially in construction where much time can be lost through disruption and interruption to work in the course of a working day. Moreover, learning curves and the division of labour also contribute to improvements in productivity, though the division of labour may also be used to control and deskill the operatives rather than only to increase productivity. Profits (after all other costs have been met) can be seen to result from the increase in output per person over and above the average wage rate paid to labour.

If productivity is taken to mean the output of a firm per person in a given period of time, it is necessary to distinguish between gross output and net output. In the construction industry with its dependence on subcontracting and prefabrication, this is particularly relevant.

Part II
Accounting for Production and Assets

4
The Logic of Accounts

Introduction

In Chapters 1–3, we saw how production processes add value to inputs using labour productivity to generate surpluses, out of which interest, rent and profits are paid. In this chapter we discuss how financial account is taken of the processes involved. We begin by looking at the accounts of the economy as a whole. These may be divided into personal, corporate and government sectors. We also make reference to the international sector.

The logic of national income accounts is reflected in our model of company accounts. Both use the same structure of production, followed by appropriation, followed by capital, and finally financial accounts. We comment on the income, output and expenditure methods of calculating national income, in order to place construction and the property sector in context and to view them in later chapters in proportion to the rest of the economy.

National accounts and company accounts in part measure the flows of income, output or expenditure per annum. We conclude this chapter by examining the relationship between these flows and the stock of wealth, and consider the relationship between flows and stocks of the built environment.

The structure of national accounts

Almost all real assets are owned either by households, firms or government. These assets represent a grand total of the stock of wealth available to an economy. Each year that stock of wealth is affected by the incomes and expenditure of its various owners. For example, when income exceeds expenditure there is an addition to the stock of wealth. Investment and saving, the differences between consumption and output, and between income and expenditure, are therefore flows into the stock of a country's assets. Investment represents additional real assets. Saving represents ownership of this stock of wealth.

Firms use their assets to produce outputs. When these outputs are sold the net revenues, distributed as wages and dividends, become the income of the household sector as well as one source of tax revenue for government. The government sector also owns real assets and produces outputs, such as health care, education services and defence. National Accounts are published annually by central government to provide information on output, income and expenditure in the whole economy.

More exactly, the three ownership sectors of the economy, to which all assets belong, are the personal sector, the corporate sector and the government sector.

The **personal sector** comprises households, businesses such as partnerships and sole proprietorships that are not corporations, private non-profit making bodies, such as charities, and finally life assurance and pension funds. Life assurance and pension funds are treated officially as part of the personal sector, because they are deemed to be simply the sum of the savings of their policy holders.

Households are the most important component of this sector. Unincorporated businesses are included in this sector essentially because it is impractical to distinguish, in the absence of limited liability, between the assets and income of a business and its owner, and because, unlike corporations, they are not legal economic entities with a separate existence from that of their owners.

Although life assurance and pension funds are financial institutions, they are included in the personal sector. Whereas other financial institutions, like banks, other insurance companies and building societies, are put in the corporate sector, life assurance and pension funds are put in the personal sector on the grounds that all their funds come from persons, their liabilities are to persons, and that, in a sense, they are merely *agents* for those persons whose pension contributions and life assurance premiums they receive. For some purposes, however, a comprehensive account, which includes life assurance and pension funds, is made for all financial institutions. As participants in the development process and property owners, life assurance and pension funds are of great significance to the construction sector.

The **corporate sector** is made up of industrial and commercial companies, financial companies and institutions, and public corporations. Financial institutions are distinguished from other companies because their accounts look fundamentally different, as they are not engaged in trading, which is buying and selling, but in lending and borrowing.

Public corporations comprise nationalised industries, plus the BBC, the Development Agencies and the like. Privatisation in the UK has meant the transfer of many large corporations out of the public corporations sub-sector, which has become rather small, and into the companies sub-sector, without in itself altering the overall size of the corporate sector.

Companies are all corporations (meaning incorporated businesses) not owned by the state. They are therefore part of the private rather than public sector. For certain purposes the whole economy can be divided into these two sectors. The private sector comprises the personal sector plus the companies sub-sector, while the public sector comprises the government sector and the public corporations sub-sector. Since the size of the public corporations sub-sector has become insignificant, much of the point of using public/private sector concepts has been lost, as the public sector has become virtually synonymous with the government sector.

The **government sector** contains central government and local authorities. Central government now consists not only of the various Ministries (like the Ministry of Defence), but also many agencies or quangos, to which functions previously performed by Ministries have been given. Quangos are quasi near government organisations, whilst agencies include the Highways Agency, the Training Services Agency, and so on. Bodies such as Hospital Trusts have a somewhat ambiguous position. They are still classed as part of the government sector rather than the personal sector, largely because the services they provide are sold to government and not to persons. Whether these Trusts own the assets they control outright, or whether government retains sufficient rights over those assets to be deemed their ultimate has been rather moot.

By definition, if an economy is closed off from the rest of the world, it is called a closed economy. However, all economies are in fact open to international trade and transactions to a greater or lesser extent. There is, therefore, a fourth sector in the national income accounts. This fourth sector, called the **international sector**, captures transactions between the three domestic sectors and the rest of the world. Although it is not a sector *within* the UK economy, it is important in the UK accounts for two reasons. First, a high proportion of output is sold abroad in the form of exports and an even higher proportion of expenditure is spent on imports. Expenditure on imports does not, of course, generate incomes in the UK, whilst exports generate UK income without there being a corresponding expenditure in the UK. Second, in an economy such as that of the UK, investment may be partially financed by overseas rather than UK savings.

To describe the whole economy as well as the built environment sector – its outputs, purchases, incomes, its real assets and its workforce – statistics are provided in a variety of annual and monthly publications. The most comprehensive source of data on the construction sector is the *Housing and Construction Statistics* produced annually by DETR. The *National Income Accounts* are also published annually, with figures including the contribution and size of the construction industry. Other useful data are given in the *Census of Production, Business Monitor,*[1] *Economic Trends* and the *Employment Gazette*. Relevant figures can also be found in non-governmental sources such as *Building* magazine, the *Architects Journal*, and Inter Company Comparisons', *Business Ratio Report*. Many detailed official statistics or production and trade are now published in CD-ROM form, for example, PACSTAT.

The different types of economic statistics available on the construction sector follow a logical but disjointed pattern. Some of the data that would complete the picture are not gathered systematically or sufficiently reliably, because they are not required by government, or because of the expense of collection or because of limited usefulness.

Figure 4.1 shows the logical pattern formed by sets of accounts. For firms, production accounts relate to income generating activities which then produce net revenues which are equivalent to gross profits. Gross profits are then distributed between corporate saving and other claimants in the appropriation accounts. Saving by firms permits investment in assets and any investment in a firm in the course of a year's trading will increase the value of assets in the capital account. As not all investment is self funded, firms need to borrow both from financial institutions and other sources. This is accounted for in the financial accounts.

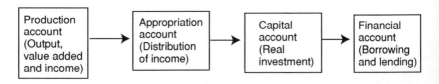

Figure 4.1   The logic of national and company accounting systems

The national income accounts also follow this pattern. Production accounts show the output of the whole economy while the appropriation accounts provide data on government, corporate and household incomes and current expenditures. The capital accounts measure investment expenditure and the asset base of the economy. The financial accounts deal with the monetary provisions needed to ensure sufficient liquidity in the system to allow it to function, and in the hands of each purchaser of output, including purchasers of real assets, to provide them with means of payment.

The logic of national accounts statistics

Government, firms and students of the economy need to know, for policy and planning purposes, and even the testing of economic theories, what was produced, how much income it generated and for whom, and, finally, where the money or income was spent. To answer these basic questions, there are three methods of calculating national economic activity, namely the output method, the income method and the expenditure method.

The **output method** measures the total produced by an economy, calculated from the net output (or value added) of all producers. Total output represents all goods and services produced in a given year. This can be expressed mathematically as:

$$Q =_{i=1} \Sigma^{i=n} V_i \qquad\qquad (4.1)$$
where

Q = national output
V_i = net output or value added in industry i.

The distribution of these goods and services depends on the distribution of income. The **income method** calculates total income from all sources, as shown in (4.2):

$$Y = W + P \qquad\qquad (4.2)$$
where

Y = national income
W = total wage income
P = total profit income.

Taxes are considered as deductions from total wage and total profit gross incomes. Income is either spent or saved and so total income also equals the sum of current expenditure and savings, out of wages, profits and taxes.

The **expenditure method** shows how the national income is used for consumption or investment. *Consumption expenditure* is concerned with the purchase of final goods and services, which are used up or consumed by their purchasers. *Investment expenditure* is used to purchase capital goods, which in turn eventually produce consumer goods. This can be expressed mathematically as:

$$D = C + I \qquad (4.3)$$

where

D = aggregate expenditure or demand

C = consumption expenditure or consumption demand

I = investment expenditure or investment demand.

Savings are actually used to pay for investment expenditure. In the case of saving by firms, this relationship between saving and investment expenditure is direct and obvious. A firm's saving is known as its retained profit. It is the residual remaining of its profit income after all current appropriations, such as profit taxes, interest payments, and dividends. Retained profit is mostly used to purchase additional fixed capital or to increase the circulating capital stock of assets. Either of these uses constitutes investment expenditure.

Savings by households is the excess of household sector income over its consumption expenditure. These savings are partly used to pay for household sector investment expenditure, chiefly, the purchase of additional new dwellings. Most household savings are placed with financial institutions. These financial institutions then lend the savings to firms to enable companies to undertake investment in excess of their own savings. Thus even the part of income that appears at first sight not to be spent, namely saving, is actually part of total expenditure, so long, that is, as savings are lent to others and not just hoarded. In any one year therefore, there is a flow of income equal to expenditure, apart from a small amount, that some individuals may hoard either at home or in safe deposit boxes.

In any given year, the value of what is produced and sold becomes income for someone, in the form of wages and gross profits. Income is either consumed or saved. Moreover, savings equals investment.

Hence, output is equal to income, which is equal to expenditure. Hence

$$Q = Y = D \qquad (4.4)$$

By definition, the three sides of the accounts must balance in a closed economy. Actual modern economies, of course, are very much open to the rest of the world. This makes it possible for national expenditure to exceed (or be less than) national income, the excess being accounted for by international borrowing or using savings made abroad. Later, as these borrowings have to be repaid, with interest, national expenditure on consumption and investment will then be less than national output, because part of that output is remitted abroad as foreign income.

Income

Firms, households and profit income

Total national income consists, in the first instance, of wages and salaries, and gross profits, as noted in Equation 4.2. Gross profits of any one firm are partly distributed as dividends and interest to the owners of its equity and its debt. These owners may be individuals, in which case they are part of the household sector or they may be other firms, in which case they are part of the corporate sector. Taking the corporate sector as a whole, payments of gross profit income by one firm to another have no effect on the overall distribution of income between sectors. We can see that profits are divided into a part retained by the corporate sector and a part distributed to certain households. Households that receive significant distributed dividend and interest income are called *rentiers*, and the payments themselves are called *rentier* income (Robinson and Eatwell, 1973).

Owners of firms and assets are called capitalists, who may be either active or passive. The former actively own and use real productive assets to create profits for their firms, while the latter own financial assets, shares or land and passively receive part of the profits made by firms. This relationship can be perceived more clearly if, for the moment, we imagine an economy without any financial institutions. Without banks, pension funds or insurance companies, equity shares and debt in firms would all be owned directly either by other firms, or by individuals.

For every real asset used by a firm to produce output there is a corresponding financial asset, either an equity share in that company or a bond of some sort representing a loan to that company. This is so by definition since any excess of the value of real assets over the debt of a firm is deemed to belong to the owners of its equity and adds to the value of this latter class of financial assets.

With the exception of *holding companies,* which conduct no business of their own but merely own shares in other companies, any industrial or commercial firm will mostly own real assets used and controlled by itself in its own production or trading activity. However, many firms also own financial assets, such as bonds and shares in other firms, which yield them interest and dividend income. In company accounts this is clearly distinguished from their gross *trading* or *operating* profit, which is the profit from their own business activity.

Acquisition of a financial asset is not the same thing at all as investment or the acquisition of real assets. All it does is transfer the use of a sum of finance in the form of money or credit to someone else, who thereby acquires a matching financial liability. This money may be used by the borrower to finance their own investment, but it may not. In any case the investment activity would then be theirs and not that of the acquirer of financial assets.

Actually, of course, in a modern economy ownership is mostly indirect. That is, individuals own stakes in pension funds and insurance policies and deposit money in banks. These financial institutions then lend money to firms, or own shares in firms. Distributed rentier-incomes are mostly directly paid by firms to financial institutions, and then by these institutions to households. In practice, therefore, the national accounts divide the corporate sector into two main subdivisions. The first consists of industrial and commercial firms, while the second consists of financial institutions.

Households and wage income

Though, as we have just seen, part of the household sector receives rentier-income, the greatest part of household sector income is wage income, which consists of wages and salaries. All wage earners are members of households, and therefore all wage income is income of the household sector.

It is worth pausing to reflect upon the economic concepts of the *household* and the *individual.* Although households may consist of families, a household exists as an economic entity whenever income and expenditure is shared between individuals who also live together, even when they are not members of the same family. Economists divide the population of individuals into the *economically active and inactive*. The former directly receive wage incomes, and the latter do not. The proportion of the population who are economically active is an important economic variable, with effects upon per capita income and living standards and upon the productive capacity of the economy. It is individuals who perform wage labour and receive wages as payment for the sale of their labour power. However, the peculiar commodity *labour power* is, at it were, produced within households. Moreover, individuals' incomes are used to sustain and benefit all members of their households, including those with no direct income of their own.

Economic relationships, such as the relationship between buyers and sellers, are assumed not to exist within a household. Instead, non-economic social relationships determine how household income is actually distributed between individuals, who decides how household income shall be spent, and by how much each individual benefits from that expenditure.

It has been suggested by some observers of modern capitalist societies that in some of them (especially, the USA) these non-economic social relations within the household, indeed within the family, are breaking down, to be replaced by explicitly economic ones – contractual regulation of the distribution of family income and wealth. Marriage itself, of course, is a legal contract, and can be interpreted by the courts to specify economic property rights of the individual parties to the contract. As divorce rates increase and inter-generational familial bonds weaken, the family becomes a more temporary and contingent economic entity than the individuals who comprise it.

In other modern capitalist societies, on the other hand (for instance, most of those of East Asia), the wider family is a very powerful and important economic unit, providing payments for the education and health care of its members. In those societies the family performs the role played by social insurance and social services in *welfare state* societies, and by private, individual savings and insurance in societies such as the USA. In this East Asian model, families may also provide the basic unit of direction and management of firms.

The concept of the household is of especial interest to students of the economics of the construction sector, because it is households, and not individuals, who determine and constitute the demand for housing, and thus generate the demand for an important part of construction industry output.

Government and taxation income

Retained profits, then, are the net income of the corporate sector, while wages, salaries and *rentier* income are the income of the household sector. Taxes are the income of the government sector. The major forms of government revenue are taxes on income, on expenditure or sales, like Value Added Tax, and on wealth, like death duties.

To understand the economic role of taxation, imagine for the moment that all government taxes take the form of taxes on income. Taxes on the income of firms, in the UK called corporation tax, and on the income of households, called income tax, then provide the government with its income. It is clear that these taxes do not create additional income, but merely redistribute, from firms and households to government, a pre-existing income derived from shares in the output and revenue of firms in the form of profits and wages.

Taxes on expenditure, like VAT, can be regarded as making deductions from the revenue of firms, before gross profit is drawn up. As such they are like costs of production, from the perspective of the firm. As economists, however, we need to note that they are actually taxes on the value added produced in a firm. Value added in the corporate sector consists of revenue from sales of output to other sectors. Value added in a single firm consists of revenue from all sales of output by that firm *less* the value of purchases of production-inputs from other firms. This value added is then divided into gross profit, wages and value-added tax. Finally, corporation tax is a tax on the residual surplus belonging to firms after all other costs have been deducted from revenues. Firms retain a proportion of net profits after corporation tax and distribute the rest to shareholders.

National income and the standard of living or economic welfare

The standard of living is often measured by *per capita* income, which is the national income divided by the size of population. Hence,

$$\textit{Per capita} \text{ income} = \frac{NY \text{ in year } t}{\text{Population in year } t} \tag{4.5}$$

Calculated annually, this measure compares the rate of growth in national income to the rate of growth in the population. As a measure of the standard of living, the major weakness of this approach is that it omits any reference to the distribution of income between the rich and

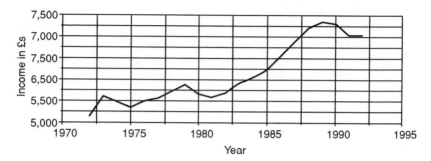

Figure 4.2 Time series of national income *per capita*, at constant 1990 prices
Source: *National Accounts* (1994 edn) (London: HMSO).

poor. Dividing national income by the size of the population implies that all members of the economy produce and receive an equal share. Nevertheless, the growth in per capita income over time demonstrates a potential improvement in economic welfare.

Figure 4.2 illustrates a time series of national income per capita. Because this is in constant price terms, inflation has been taken out of the figures. Constant price data can, in principle, be used to represent change in the purchasing power of income over time. Purchasing power is the size and content of the basket of goods an income can buy. From Figure 4.2 it can be seen that purchasing power grew throughout the period, though not in every year.

Output

Output is a flow of goods and services stemming from the production process. When the output is sold the cost of inputs is deducted and the remainder is then distributed as income to the workers, management and owners. Hence, output can be measured in terms of the value added to inputs by firms and this valued added is the source of all income. National income measures output. However, it is important to bear in mind that not all income is directly related to output. Transfer income is income received where there is no corresponding output. Pensions, welfare benefits and gifts are all examples of transfer payments and these must not be included in measuring the value of the output of an economy. Transfer payments are a method of distributing output to those who would otherwise receive a smaller share, in much the same way as a child's pocket money paid by a parent does not increase the income of a household but only enables the child to spend some of the family's income.

As we have seen in Chapter 1, this value added approach can be adopted to analyse the construction sector. It is possible to follow raw materials from their extraction from nature, through the production process, to their finished state. Economists call this transformation a vertical progression, as each firm adds value to the raw materials at each stage of production before passing them down the line towards completion.

Domestic and national product

The total value added or net output of all firms in an economy is called the Gross Domestic Product (GDP). The GDP estimate is based on the returns which firms provide. These returns are based on selling prices, which relate to new goods, which immediately begin to depreciate. The term *gross* is used to note that when calculated, no allowance has been made for this or any other depreciation. The term *domestic* refers only to output or *product* made within the national borders of an economy.

Gross National Product (GNP), by contrast, includes income made abroad by firms and individuals who are citizens of a particular country. However, it also deducts income belonging to foreign citizens and remitted abroad, although made in that country. These two adjustments to GDP are combined to give a figure called the net property income from abroad (*NPIFA*). Thus

$$GDP + NPIFA = GNP \qquad (4.6)$$

Note that no attempt is made in the UK national accounts to measure the total net output (value added) of these foreign activities, but only that part of the output resulting in incomes for UK nationals, the repatriated profits and wages. These modifications to the GNP are really 'income' and not 'output' adjustments. The concepts of 'national' and 'domestic' *income* make clear sense.

'Output' from real estate

GDP at market prices is equal to total sales receipts at over-the-counter prices less bought-in inputs. This is equivalent to total value added at market prices. However, it is also necessary to include the value of the annual imputed output from buildings, measured in actual rent paid or in imputed rent in the case of owner occupation. Jackson (1982) points out that if the rent for a factory, office or shop is actually paid to a landlord, then the output of that property is included in 'property output'. However, the imputed output of owner occupied buildings, except for dwellings and farms, is included in the value added of the industry or sector that owns them.

Measuring GDP (or, measuring economic aggregates)

(a) Current prices or constant prices

When national accounts statistics are gathered firms report their current expenditures and sales. The growth in the size of national output as measured by these economic statistics at current prices is partly due to inflation and partly due to the growth in the size of economic activity. There is therefore a need to distinguish real or actual increases from increases caused by higher prices, because only the former indicate an increase in total economic welfare or well-being. For some purposes, however, it is the current money value of GDP that is relevant. In any case, this is the form in which the figures first arrive.

The National Accounts are given in both current and constant prices. Constant prices are an attempt to remove the effects of price inflation and provide estimates of the volume or amount of output of all firms.

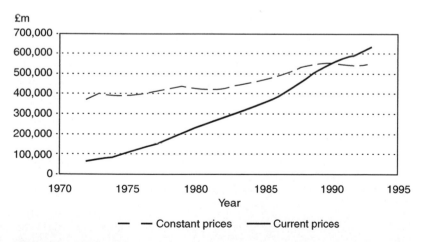

Figure 4.3   Time series of GDP in current money and constant 1990 value or 'volume' terms, both at market prices
Source: *National Accounts* (1994 edn) (London: HMSO).

Constant prices use the prices in a particular year to calculate over time the rise in national income caused by inflation. A real increase in national income occurs if the percentage increase in national income is greater than the percentage increase in prices.

Figure 4.3 shows how GDP has grown since 1972 in both volume and money terms. It can be seen that while the monetary valuation of GDP has grown rapidly in the period, due largely to inflation, the real growth of output has been considerably slower.

(b) Gross and net product

Economic aggregates such as gross domestic or gross national product provide values before depreciation has been taken into account. Gross domestic product is already, in one sense, a measure of net output. It already removes double counting of the value of inputs of materials and the like. Gross output minus the value of purchases of inputs of materials, components, services and energy costs is called net output. As mentioned earlier in this chapter, the sum of net outputs is called the gross product. It is 'gross' because there is another deduction still to be made, that for use of fixed capital inputs.

Net product is the value of gross product less the value of inputs of fixed capital used up, consumed or worn out in the course of production. Our main way of measuring this consumption of fixed capital is through depreciation. It is a cost of using or consuming durable goods such as vehicles, machinery and buildings and is referred to as 'capital consumption' in the national income accounts.

An accurate measure of the value of output available as income must take depreciation into account. Deducting the value of depreciation from the GNP or GDP figures gives the Net National Product (NNP), and Net Domestic Product (NDP), respectively. Examples of these measures of national income are given in the Table 4.1.

Table 4.1   National income, at current factor cost, 1983–93

Measure of national income (£m)	1983	1988	1993
GNP	264055	405852	549182
NNP	227905	353216	484159
GDP	261225	401428	546120
NDP	225075	348792	481097

Source: *National Accounts* (1994 edn) (London: HMSO).

Accuracy and reliability of the components of the National Accounts

The national income accounts record figures reported by government, firms and individuals. Self-employment incomes and incomes from the 'black economy' have to be estimated based on assumptions about the degree of under-reporting of incomes, and the amount of unrecorded output, factors which have to be noted in judging the accuracy of the estimates. Similarly, favours and unpaid work including domestic chores and do-it-yourself repairs are excluded from the estimates of national income although these activities all contribute to the flow of goods and services produced in a given year.

Further complicating factors arise, for example, because households receive important benefits in kind, chiefly, free use of public services, that are like income but do not appear as part of their money incomes.

The imputed incomes and expenditures are, of course, the items subject to greatest inaccuracy of measurement. Capital stock values are also uncertain, because they depend so much on assumptions about the lifespan of items of capital.

Expenditure

The proportion of expenditure between governments, firms and house-holds varies from country to country depending on their stage and pace of development. It is not possible to generalise for all countries or economies. Once an economy is no longer rapidly industrialising or urbanising it may be said to have reached a mature phase. As a crude approximation, in most mature capitalist economies, total expenditure for recent years divides roughly 60:20:20 between personal consumption, investment and government consumption.

Though investment is only around 20 per cent of aggregate expenditure, we will concentrate most of our attention on this component of total demand. One general reason for doing this, in any work concerned with the whole economy, is that it is fluctuations in this investment expenditure that lie behind, and are the immediate cause of, most of the fluctuations in total demand. Hence fluctuations in investment demand are the cause of periods of recovery and periods of recession in the business cycle. There is a further, more specific, reason for this focus when our concern, as here, is with the construction sector, viz. that the greater part of all construction demand consists of investment demand.

The structure of company accounts

In principle, the national accounts follow a similar overall pattern to the accounts of a company. This is important, because national accounts, on the one hand, and company accounts, on the other, are our two main data sources, as industrial economists. It is important that we are able to understand the relation between the one and the other – their fundamental similarities, but also their differences.

Many of the differences arise from the principle of composition. The national accounts are based upon this principle. They compose or treat as one all the firms or individuals in an ownership sector or industry. To give an example, suppose some households (call them Type A) are savers (income > consumption expenditure), whilst others (Type B) are borrowers or dissavers (income < consumption expenditure). The national accounts report a single figure for net household sector saving, the excess of Type A households' savings over Type B's borrowing or dissaving. The implicit assumption is that it is 'as if' the B households borrowed what they needed from the A households. Transactions within a sector are then disregarded. This is called *netting out*. Likewise, if ownership of an asset is transferred from one firm to another, this is ignored in the sectoral accounts. It is only if all firms, taken as a whole, buy more assets than they sell to one another that an item for acquisition of assets will appear in the sectoral accounts. To a firm, on the other hand, all other entities with whom it engages are the outside world, and are treated in the same way in the company accounts, regardless of whether or not they are in the same sector or industry.

In passing, we should also note the existence of the fallacy of composition. This means that it is often an error to reason that what is true for one individual, household or firm will therefore, by extension, also be true for all individuals taken as a whole. For example, if one firm reduces its wage bill whilst leaving the value of its output unchanged, its profits will increase. However, if all firms reduce their wage bills by employing fewer workers or paying lower wages, it is by no means the case that their profits will increase. If all workers have lower incomes, total consumption spending will fall, and firms collectively will find themselves with some unsold output as a result. What is true for one firm acting alone, does not apply if all firms behave in the same way. As we move to and fro from discussion of micro- to macro-economics, from the single economic entity to the aggregate and back again, we will have to be aware continuously of this kind of problem.

All accounts are either stock or flow accounts. Stock accounts usually describe stocks of wealth at a moment in time. Flow accounts describe flows of revenue and outgoings, of income, of saving and investment, of changes to the stock of ownership of financial assets, over a period of time, usually one year. In company accounts, the balance sheet is a stock account, and the profit and loss account and sources and uses of funds or cash flow statement are flow accounts. Similarly, in the national accounts, there are stock accounts for capital and for financial assets, whilst the rest are flow accounts.

We are now in a position to understand the logic of company accounts as well as national accounts. Both sets of figures consist conceptually of four sets of accounts, concerned with production, expenditure, capital and finance.

The production account

We begin with the value added or production account, which shows how income is derived. The value of sales is the value of gross output, from which the purchases from other firms are deducted. The final item of the account is in theory, therefore, the aggregate value added in a given year. This fits in with the national income accounts, which sum value added in each industry to produce the output method GDP. An industry's value added is itself the sum of the added values of all the firms comprising that industry.

In company accounts, at least in the UK, value added is not directly reported. Instead the production or operating account starts with sales revenue or turnover and deducts the *cost of sales*, or all the direct costs of materials, services *and labour* used in producing the output sold, to arrive at the firm's gross operating profit. From this, 'other expenses' (primarily management overheads and depreciation) are deducted to arrive at net operating profit. Taking figures from the table below, it can be seen that Tarmac's[2] net operating profit in 1996 was £117.5 million.[3] Thus wage payments are included with the cost of purchases from other firms. Sometimes a note to the accounts states the value of the wage bill, enabling a figure for value added to be calculated, but firms are not obliged to provide this information. Tarmac however, published its total wage cost at £523.1 million, which, added to the net operating profit, and adding back in depreciation £68.5 million, gives a figure for gross value added of £708.1 million.

When wage costs are not given it is to some extent possible to overcome the difficulty of finding value added by turning to the Census of

Extract from Tarmac Group Profit and loss Account for the year ended 31 December 1996

	£m
Turnover	2,663.8
Less cost of sales	(2,183.6)
Gross operating profit	480.2
Less other expenses	(362.7)
Operating profit	117.5
Less exceptional non-operating items	(65.0)
Profit before interest	52.5
Net interest	(42.0)
Profit before taxation	10.5
Less taxation	(2.9)
Profit afetr taxation	7.6
Less minority interests	(4.0)
Profit for the financial year	3.6
Dividends	(50.9)
Transfer from reserves	(47.3)

Notes to accounts as given in the annual report:

From note 12 Depreciation of tangible assets	68.5
From note 7 Total wage costs	523.1

Figure 4.4　Profit and loss account
Source: Tarmac plc Annual Report, 1997.

Production (COP), which collects data concerned with production industries. This published data summarises returns made by a sample of firms in each industry. From it we can see the composition of value added for each industry, i.e. the ratio of wages to profits. This can be compared with the income-method breakdown of industry value added in the national income accounts.[4] Each firm surveyed for the COP reports its gross output, its value of purchases from other firms, its value added, its wage bill and its gross profits, amongst other things.

The appropriation account

Having derived income from the production set of accounts, the next step is to account for its distribution. This can be found in the income and expenditure or appropriation account. This account, whether in the national income accounts or company accounts, starts with operating profits. From this, we deduct interest payments, profit taxes and dividends to arrive at undistributed profit, sometimes referred to as corporate saving. In company accounts this final figure is called retained profit, while in the national income accounts, it is called corporate sector saving. The two would be essentially similar, except for the fact that some interest and dividend payments made by firms are received by other firms. Thus these payments by some firms to other companies do not appear as a deduction from the gross profits of the corporate sector in the national income accounts. However, in the individual company accounts, these payments to other firms are treated in the same way as payments of dividends to those outside the sector, such as private individual shareholders. All interest and dividend payments must be deducted to arrive at retained profit.

In the Tarmac Group accounts for 1996, profit income consists of two flows. First, and normally by far the largest for an industrial or commercial firm, is the operating profit, derived from the production account at £117.5 million. But additionally there are non-operating flows of profit income. These income flows stem from the firm's secondary *rentier* role, as a shareholder in other businesses or as a lender to other firms or to government. In Tarmac's case, there were *losses* on exceptional non-operating items of £65 million, which reduced profits before interest to £52.5 million. Inter-firm payments of dividends and interest appear as deductions from total profit in the company accounts of one firm and as an inflow or contribution to total profit in the company accounts of other firms. These inter-company flows are included as 'other income'. Thus some firms' retained profits are thereby increased by as much as others' are reduced, and the flow is netted out in the national income accounts, as mentioned earlier.

In company accounts, a distinction is always made between interest and tax payments, on the one hand, and dividend payments on the other. This is because, at least in principle, the firm belongs to its shareholders, and the purpose of company accounts is to account to the shareholders for the use the firm has made of their capital, and to show how much return shareholders have made on that capital. This shareholders' return consists of two elements; the dividend income they receive (in Tarmac's case £50.9 million), and the increase in the value of the net assets of the firm, as a result of investing retained profits to expand the assets of the business. Unfortunately in 1996, the value of the net assets of Tarmac were not increased but reduced by £47.3 million which was transferred from reserves to cover interest, taxes, losses on exceptional non-operating items involved in the company's restructuring and dividends. Net profit is the profit after deducting interest and tax payments, but before payment of dividends. In 1996, Tarmac's accounts show the company made a net profit of £3.6 million. Net profit is therefore a measure of the total return to shareholders. While net profit is vitally important in company accounts, as a concept it plays a negligible role in the national income accounts.

The capital account

The capital account illustrates in monetary terms changes in the stock of capital caused by inward flows of net capital additions or outward flows of capital consumption. The inflow to start this account is retained profits or, more generally, saving, which is the last item in the previous appropriation account.

Retained profits appear in the company sector in the national income accounts, as well as in individual company accounts. Most investment expenditure is financed from retained profits or company savings. It is investment which has the effect of increasing the stock of fixed or circulating capital. Investment is therefore often referred to as capital formation. Both company and national income accounts are careful to describe capital formation in some detail, distinguishing between gross and net capital formation. Gross capital formation is simply the total of investment expenditure in the year, while net capital formation is calculated by deducting capital consumption or depreciation from gross capital formation.

Allowing for depreciation reduces the remaining value of fixed capital assets as they get older. This provision for depreciation is handled in a special way in company accounts. It is treated as if it is a cost, and is deducted from revenue to arrive at operating profit. In Tarmac's case, depreciation of fixed assets in 1996 amounted to £68.5 million. This provision for depreciation reduces a firm's reported pre-tax profits and means that payment of tax on profit income is thereby reduced by the amount of depreciation multiplied by the percentage tax rate. However, this provision is neither an actual cost nor is it an outgoing. It is purely notional as no cash payments are made. This part of revenue remains in the hands of the company. Gross retentions of a firm equal retained profits plus depreciation provisions.

Extract from the Tarmac Group Balance Sheet at 31 December 1996

	£million	£million
Assets employed:		
Total fixed assets		1440.3
Current assets	1,163.5	
Less Creditors (current liabilities)	1,058.5	
Net current assets		105.0
Total assets less current liabilities		1543.3
Long term liabilities (due after more than one year)		683.0
Net assets		862.3
Financed by:		
Shareholders funds	778.7	
Minority interests	83.6	
Total capital employed.		862.3

Figure 4.5   Balance Sheet
Source: Tarmac plc Annual Report, 1997.

Reflecting the restructuring arrangements with Wimpey and continuing difficult trading conditions in the construction industry in 1996, Tarmac was unable to retain any profits and therefore did not increase its capital base. On the contrary, in order to meet the income appropriations as noted above, it was obliged to reduce its reserves by £47.3 million. The overall picture revealed in the accounts shows that Tarmac's retained profits in 1996 were negative, amounting to a loss of £47.3 million from reserves. Depreciation provisions were £68.5 million. Thus the final figure for gross retentions for 1996 was a surplus of £21.2 million. But depreciation reduces the book value of the firm's (fixed) assets, and thus net change in the value of assets (capital employed) was negative.

In this balance sheet, economically the most significant figure is 'total assets less current liabilities'. This represents the sum of the long term debt and equity of the group, and is what economists normally have in mind when they talk about the capital finance of a firm. The ratio of long term debt to equity is known as the gearing ratio.

The financial account

Finally, there is the financial account. In everyday speech, no clear distinction is made between acquisition of real or tangible assets and financial assets. Both are loosely called *investment*. In economics, however, we reserve the term investment for the acquisition of additional real assets, which add to the total capital stock and productive capacity of the economy.

If real investment in a sector or a firm does not equal saving, then there is a corresponding financial surplus or deficit. A financial surplus results if savings exceed investment. The firm or sector can then use the financial surplus either to acquire ownership of additional financial assets, or reduce its previous financial liabilities. A financial deficit, on the other hand, means that the sector or firm has financed part of its investment by issuing new financial liabilities, i.e. by borrowing or, as occurred in Tarmac's 1996 accounts, by selling some previously held financial assets, such as shares in subsidiaries or associate companies. Thus, from the firm's cash flow

Extract from the Tarmac Group cash flow statement for the year ended 31 December 1996		
	£m	£m
Net cash inflow from operating activities		59.5
Net cash outflow from investments and interest		(46.3)
Corporate taxation		2.4
Investing activities		
Purchase of tangible and fixed assets	(76.2)	
Sale of tangible and fixed assets	15.7	
		(60.5)
Exchange of business with George Wimpey plc	45.3	
Purchase of subsidiary undertakings	(40.9)	
Sale of subsidiary undertakings	43.6	
Net cash inflow from investing activities		48.0
Equity dividends paid		(49.1)
Net cash inflow before financing		(46.0)
Financing		
Issue of ordinary shares	1.1	
Short and long term bank loans	77.4	
Net cash inflow from financing		78.5
Increase in cash		32.5

Figure 4.6   Cash flow statement
Source: Tarmac plc Annual Report, 1997

statement below, it can be seen that net investment in tangible and fixed assets was £60.5 million which was financed by the net disposal of financial assets in subsidiary undertakings and the exchange of business with another contractor of £48 million, as well as issuing shares during the year worth £1.1 million and borrowing £77.4 million. This use of external finance enabled Tarmac not only to pay its shareholders a dividend of £49.1 million, but also left it with an increase in its cash reserves of £32.5 million at the end of the year.

One entity's acquisition of a financial asset is balanced by someone else's equal acquisition of a financial liability. The very same piece of paper, such as a bond, which is an asset to its owner is a liability to its issuer. This is a reshuffling of claims on national income. Financial transactions redistribute the immediate use of money wealth from savers and lenders to borrowers. Debt finance implies a reverse redistribution of income in the future from borrowers to lenders. In the process of these financial dealings, nothing is done to raise productive capacity or future national income. Financial assets are merely, in effect, claims to a share in whatever gross income may be made by the entity acquiring the financial liability. Should this income prove insufficient to meet these claims, the owners of financial assets then usually have the right to the proceeds from the forced disposal of the real assets of the borrower.

Thus real assets, owned by the borrower, need to underpin the financial assets owned by the lender. Most lending in a capitalist economy is to businesses. Therefore, interest is a claim on gross profit incomes. Consumer credit to wage earners means that credit interest payments also become a claim on future wage incomes.

The typical form of financial asset and liability is indeed the loan. A bank balance created by a bank granting a loan adds a corresponding financial liability into the accounts of the borrower, and a financial asset into the accounts of the bank. The borrower is liable for the repayment of the debt. A company depositing money with its bank involves the reverse – the firm acquires a financial asset, and the bank a financial liability. In this instance, the bank acquires a debt corresponding to the size of the deposit.

Apart from loans, bonds and other kinds of debt, shares in companies are another form of financial asset. Although shares are a financial asset to the share owner, they are not a legal liability to the company that issued the shares, in the sense that the issuing company has no liability to repay the nominal value of the shares although there is a

virtual liability to pay dividends. Hence, in spite of incurring a reduction in the value of its net assets from £928 million in 1995 to £862
million in 1996, the directors of Tarmac decided that the firm should
pay its shareholders a dividend of £50.9 million.

National accounts again – the composition of national income

In Table 4.2b, income from self-employment, gross trading profits and
rent are all shown before providing for depreciation (i.e. are gross of
depreciation). NNP, on the other hand, shows what national income
would be after providing for depreciation – that is why it is called 'net'
national product. Total net income from self-employment, gross profits
and rent would therefore be £ billion £107 billion (i.e. 57 + 62 + 50 – 62);
the sum of incomes to owners of capital less provision for depreciation
('consumption') of that capital.

To measure income *gross* in Table 4.2b, we have to add back any
income 'that would be available if no deduction were made for depreciation'. This is already done for companies, in gross profits. The last row

Table 4.2a   National accounts, 1992

	£ billion	
	Sub-totals	Totals
GNP at market prices		601
Net property income from abroad	4	
GDP at market prices		597
Factor cost adjustment (taxes on expenditure less subsidies)	81	
GNP at factor cost		520
Capital consumption	62	
NNP at factor cost		458
GDP at factor cost		516

Table 4.2b   National accounts, 1992, by type of income

	£ billion
Income from employment	343
Income from self-employment	57
Gross trading profits	62
Rent	50
Imputed consumption of non-trading capital	4
GDP at factor cost	516

Table 4.2c National accounts, 1992, by sector of employment

	£ billion
Personal sector income	137
Corporate sector income	294
Government sector income	85
GDP at factor cost	516

Source: National Accounts, 1995 edition, Tables 1.1, 1.4 and 2.6.

makes the parallel allowance for the other, 'non-trading' sectors, i.e. government and non-profit making bodies. Note its small size. We shall have cause to return to the question of the rate of depreciation on government-sector built stock later, and discuss why the figure shown above is so low.

The figures in Table 4.2c are the *value added* figures on the production account for each sector. That is, they are the sum of wage incomes of people working in a sector plus the gross profit, gross trading surplus, rent and self-employment incomes of businesses and other organisations in a sector. Thus, for instance, the figure for personal sector income is not the income that *ends up* in the personal sector, but the income generated by *production activities* regarded as taking place within the personal sector, i.e. in unincorporated businesses or by the self-employed.

Income and expenditure in the corporate sector

In the *Blue Book* (1993) (Chapter 5 – Companies sector accounts) we can likewise trace gross profit income through the appropriation and capital accounts, to arrive at the financial balance.

In Table 4.3 the figure for non-trading income arising in the UK refers to the whole corporate sector including financial companies. The profit income of financial companies is viewed as non-trading income.

It is clear, then, that by producing output, firms add value to their inputs. The value added is then taxed and divided between the owners of firms, the owners of land and buildings used, the owners of finance borrowed, and the labour employed by the firms. Their income net of tax can then be saved or used to purchase goods and services, while government can use tax revenues for public sector spending. At the same time, firms use their retained profits and

Table 4.3   Financial balance of the corporate sector, 1992

Corporate sector	Sub-totals £ billion	£ billion	Totals £ billion
Appropriation account			
Gross profit income			147
arising in the UK			
gross trading profits made in UK	65		
rent arising in UK	5		
non-trading income arising in UK	49		
sub-total gross income arising in UK		119	
arising abroad			
profit on direct investments abroad	14		
other profit arising abroad	14		
sub-total gross income arising abroad		28	
appropriation of income			(99)
dividends	24		
interest payments	53		
remitted abroad	5		
UK taxes on income	15		
Other	1		
Balance undistributed profit income			47
Capital account			
Gross retentions (receipts)			
Undistributed income			47
less provision for stock appreciation	–2		
less provision for depreciation	–36		
after providing for stock appreciation and for depreciation		9	
Investment (expenditure)			(54)
Gross fixed capital formation		54	
of which			
replacement of capital consumption	36		
net fixed capital formation	17		
increase in value of circulating capital	0		
Balance financial deficit (gross retentions *less* investment expenditure)			(6)

borrowing to maintain and increase their productive capacity over time.

It is important to distinguish between the mass of profit and the rate of profit. The mass of profit is the total amount of cash generated remaining after all other costs have been met. It is the total of all distributed and retained profits. The rate of profit is the amount of profit per £1 of capital owned and used. It is therefore quite possible for a large firm to generate a larger mass of profit than a smaller firm which generates a higher rate of profit.

The ability of firms to invest depends on the amount of profits they retain, which is derived from the mass of past profits, and the amount of money they borrow, which is affected by interest rates and the climate of business confidence. The last factor is important because investment is essentially concerned with the future, and the firm's expected ability to repay any loans in the future. The confidence firms have to invest therefore depends on the expected rate of profit needed to cover the rate of interest and repay a loan.

Retained profits are a major source of internally generated finance. Moreover, the greater the amount of retained profits, the greater the assets of the firm and, as lenders gear their funding to the assets of the borrower, the more money the firm can borrow. As an alternative to borrowing, firms may issue shares, which enables joint stock companies to expand their monetary assets by drawing in more shareholders. This dilutes the control of existing shareholders in the hope that the new funding will allow the firm to expand.

Investment demand fluctuates from year to year depending on past performance and expectations about future market trends. As most major investment decisions involve construction, fluctuations in demand for construction follow the pattern of investment decisions. As demand fluctuates so too does construction output. Thus, a kind of dynamic economic cycle is created. Profit income partially determines investment expenditure, which in turn determines output, and output in turn determines profit income. This accounts for fluctuations in construction output from year to year.

However, taking a longer term perspective, say over decades, these fluctuations begin to appear as fluctuations in the rate of growth in the stock of the built environment over time. Investment in buildings is measured in the national accounts as part of net capital formation. As this investment is durable, it tends to be cumulative as the new building work in any one year is additional to the existing stock of buildings and infrastructure, after demolition and excess of capital ageing over repair and maintenance have been taken into account. In this important sense construction as a part of net capital formation is a major contributor to the rate of growth in the productive capacity of the economy.

Stocks and flows of the construction industry

Having looked at the stocks and flows in the accounts of the national economy and an individual firm, we now turn to the stocks and flows of the built environment sector. Figure 4.7 shows the relationship between the stock of the built environment and the flow of construction activities which cause the stock to increase over time.

In Figure 4.7, t_0 represents the start of year 1, and t_1 is the end of year 1 and the start of year 2. The flow chart begins with the stock of the built environment at the start of year 1. This stock provides the potential services of the built environment, measured in neoclassical economics by the sum of rents and imputed rents for the use of all buildings and the imputed value of the use of all of the built infrastructure. The potential flow of services depends on the quantity of existing buildings and infrastructure. Each year a certain number of buildings

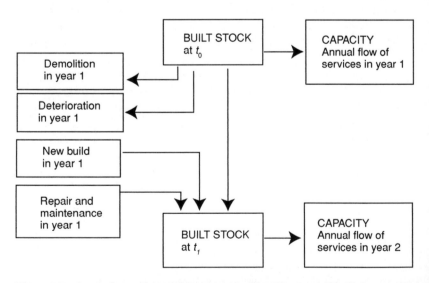

Figure 4.7   Stock flow relationships in a world without qualitative economic change: the gross capital stock

are demolished. While the remaining stock gradually deteriorates with time, some of it receives spending in the form of repair and maintenance, whereas the rest continues to serve without any intervention by construction firms. Moreover, each year some new buildings and civil engineering works are completed and they are added to the built stock. At the end of the year the total stock of buildings and structures provides the capacity to produce a flow of services is the following period. Of course this is a schematic simplification, since stock is continually changing throughout any year and consequently this alters the capacity of the built environment, continuously, rather than in discrete annual jumps.

Capacity of the built stock is a measure of the potential of that stock to provide a flow of services in the form of useful accommodation of activities. Gross capital stock is a measure of that capacity, in contrast to net capital stock, which is a measure of the book value of built stock after depreciation, which is given in the accounts of its owners. In a simple model of stock replacement, we can begin by assuming no technical or qualitative demand change. This permits the use of valuation at replacement cost. Inflation would then be the only reason replacement cost did not equal historic cost. The net value of an item of capital stock can be computed by combining a measure of replacement cost with a measure of depreciation, and this may or may not be approximately equal to its disposal or market value. In company accounts, firms mostly value capital stocks at historic cost less depreciation. However, this approach begs the question that there is a fundamental potential difference between a valuation based on historic or replacement cost less depreciation and a valuation based on the value from expected profitability over an asset's remaining economic life.

This difference explains why both profits and losses arise on disposal of assets. Disposal value does not equal book value. Market value does not equal historic cost less depreciation. This is also the reason firms revalue their assets up or down from time to time. It also creates a fundamental difference between the valuations of net fixed assets in company accounts and the Blue Book's valuations of net capital stock, which are put at written down replacement cost. This is done simply so that annual capital consumption is measured in the same prices as current transactions, and thus can be sensibly deducted from GDP to compute NDP or national income (*UK National Accounts, Sources and Methods*, 1985 edn, p. 199).

Now, in a world of no qualitative change in demand or technology, we could make use of the CSO/ONS concept of net capital stock at replacement cost. In such a world, retired assets would actually be replaced by identical equivalents. We could therefore value existing assets by finding the market value of an identical, new replacement asset and then deducting depreciation from this, to reflect the fact that the existing asset is not in fact new, and does not have as many remaining years of economic life in it as a new identical asset would.

The capital stock could then only change in two dimensions. First it could grow or shrink in terms of the number of assets of a constant type in existence. Second, it could age or get younger in terms of the average age of assets in the existing stock. However, in a world with qualitative economic change, two things happen to add two more dimensions to the possibilities for change in the capital stock. Existing assets can be *modified* to adapt them to meet new demands or embody new techniques of production, and new assets are added to the stock which are *not* identical equivalents of existing assets. This is the real world of qualitative economic change that we will try to deal with in the next chapter.

In Figure 4.8 the total stock consists of all the sub-stocks combined within the area bounded by the dotted line. Each existing sub-stock ends the year with a lower value than it had at the start of the year. This is shown by the depreciation outflows. With straight line depreci-

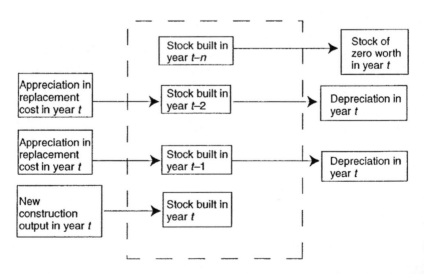

Figure 4.8 Stock flow relationships in a world with no qualitative economic change: the net built capital stock

ation, these are each equal to 1/n th of their initial valuations. However, if prices of capital goods are rising in current price terms through time, this fall in value is offset by the *appreciation of replacement cost* inflows.

Whereas repair and maintenance can be made to fit into a modified *gross* capital stock diagram, by making the assumption that there will be a natural decline in productive capacity of an asset as it ages, and deteriorates, and that this can be offset or countered by repair and maintenance activity, it is hard indeed to fit repair and maintenance activity into a *net* capital stock model. Really, this can only be done by allowing the lifespan of an asset to be a variable rather than a predetermined constant. Then the greater the flow of repair and maintenance the longer the assumed asset lives, the smaller the annual depreciation and the greater the duration, n, before stock is fully depreciated so that its value is zero and it makes no contribution to the valuation of the whole net capital stock. However, the whole net capital stock valuation method is based on the idea that assets have a predetermined fixed lifespan. Only thus can the perpetual inventory method be used to convert data on fixed capital formation into estimates of value of net capital stock. Remove this assumption and we would need a method of directly measuring capital consumption per year and no such method exists.

Concluding remarks

Taking a set of company accounts, one can model the data according to a logical sequence of events. The first stage is the production phase and this is measured in company accounts using the operating account. Analysis of the operating account reveals the value added by the firm in the course of trading over a year. Having produced added value in the course of a year's trading, profits (or losses) are created. The appropriation of these profits between interest payments, rent, retained and distributed profits can be seen in the profit and loss account. The balance sheet can then also be used to show changes in assets (or capital) reflecting investment decisions in the capital account. Finally, borrowing and lending is revealed in the financial account based on the balance sheet and the sources and application of funds statement of the firm. National accounts can be used to model the same production, appropriation, capital, and financial accounts at an aggregate national level.

If one attempts to apply flow and stock concepts to the built environment, it is clear that one measure of the built environment would need to be based on its capacity. Such a method enables aggregation of the diverse buildings and structures that comprise the built stock. However, changes are continually occurring to that built stock. This is the work of the construction industry, like gardeners tending their plants. As structures are demolished or retired the quantity of the built environment is reduced, while construction work on existing stock and new construction maintains and enlarges it.

The next chapter deals entirely with stocks through time. Statistics of stocks provide information on the quantity of accumulated resources on any given day. In the real economy, the main economic stocks are fixed and circulating capital, as well as the workforce, while in the money economy, the main economic stocks consist of financial assets and liabilities.

Notes

1. *Census of Production* and *Business Monitor* were replaced in 1996 by PACSTAT CD-ROM, and *Census of Production* renamed *Annual Survey of Production and Construction*.
2. Tarmac plc, one of the UK's largest construction sector corporations; subsequently divided into two corporations, Tarmac and Carillon plc.
3. Firms usually deduct depreciation after stating gross operating profit. That is, it is included with administration overheads in 'other expenses'. However, it can be deducted earlier, as in Tarmac's case. In any event, we need to remember that depreciation is not an actual expenditure, and that gross valued added should be calculated *before* deducting depreciation.
4. One quirk, that causes much difficulty in analysing the construction industry in particular, is that, in the national accounts, self-employment incomes are grouped together with profit income. The rationale for this is the idea that self-employed individuals are in effect petty-capitalists, and that at least part of their income consists of 'profits', whilst the other part consists of the 'notional wage' they would have earned as workers had they in fact sold their labour-power to an employer instead of 'working for themselves'. Since this income is, conceptually, a mix of wage and profit, it might as well be put in the one category as in the other. This is valid for 'jobbing builders', self-employed individuals running businesses and selling their output to customers. However, most 'self-employment incomes' in construction are not of this kind, but are the wage incomes of 'labour-only subcontractors' (LOSC) and properly should be transferred to the wage income figure. We are not able to make this adjustment because we do not have regular data on just what proportion of all 'self-employment' income in construction consists of LOSC wages.

5
Value of Stock and Flows of Built Capital at National Level

Introduction

This chapter deals with the central issue of capital, its definition, and valuation. We look at the problem of the valuation of land, and discuss various concepts of capital. Capital is linked to the productive capacity of an economy, via investment in fixed capital, especially buildings and infrastructure. The chapter ends by applying the stock and flow concepts of Chapter 4 to capital and investment in the national income accounts.

Built structures, whether buildings or civil engineering works, take substantial time to produce. Once produced they are consumed or gradually used up over the longest periods of any produced commodity. With certain levels of expenditure on repair and maintenance they need never be used up at all, in the sense of becoming physically worn out. However, it is more doubtful whether any level of expenditure on improvement and adaptation of buildings can necessarily indefinitely postpone the moment when eventually they no longer meet a need. The land on which built structures sit, on the other hand, is not produced at all, nor, in principle, is it ever necessarily used up.

The valuation of property and real estate

Several alternative methods are commonly used for valuing property. In one, the capital value of a piece of property depends on the expected future income it will provide for its owner. This income in turn depends on the expected revenue from rent less the expected cost of maintaining and operating the property. The expected revenue of a building or land depends upon its expected usefulness to future occupants. This method of valuing property can be applied to buildings whether or not they are owned by the occupier. Owner-occupiers are regarded as paying themselves an imputed rent, as if they were their own landlord. The multiple of next year's net rent that gives the capital value is called the *years' purchase*. Alternatively, *all* future years' net rents can be discounted to their equivalent present value to give a net present value (NPV) version of capital value (Gruneberg, 1997).

Ownership of the built environment normally takes the form of a combination of buildings and land. While the built structures themselves normally lose their capital value as time passes, the land or site often gains or appreciates in capital value over time. The combined capital value of property may rise or fall over time depending on which process is the stronger – the depreciation of the built structure or the appreciation of the land.

Not all built structures have capital values derived in the same way as privately owned property. For instance, many buildings and built structures are public goods, such as roads and bridges, the use of which is supplied free of charge to users. The value of any given built structure as a form of capital does not depend upon its physical characteristics or type of function or even its value to users. The value of property as a capital asset depends on who owns it and whether or not the usefulness of the structure contributes to an output that is in turn sold or commodified. Its value also depends on whether or not the land and the buildings can be sold together or separately without restriction and whether or not the site or the buildings can be converted to other uses.

If there are no restrictions on its sale, a property is said to be freely *alienable*. That is, it is possible to separate the rights to use a property from the rights of the general public. For example, if land presently owned by the state and used for a park or a school is alienable and convertible then it has a potential capital value equivalent to its opportunity cost, that is, its value in the highest rental paying alternative use. Since the early 1980s, an increasing proportion of the built stock in the UK has been given a capital value in this way, mainly because of infrastructure and housing privatisation and the increased alienability of property, which had originally been built to provide public goods.

For built stock which is not commodified, then, because it has no realistic rental value and hence cannot have a capital value derived by capitalising this rent, alternative procedures for valuing property are used to measure its value in the accounts of its owner. These valuation methods are based either on original historic purchase prices or upon replacement costs, less an allowance for depreciation. These indeed are the methods used to value the entire built stock in the National Accounts.

Historic prices and replacement costs can, however, produce valuations that are very different from those based upon expected net income from the ownership of the property. Indeed, land is not valued at all in the capital formation and capital consumption parts of the National Accounts and is not even counted as a part of the national gross capital stock, because it is neither produced nor destroyed, and therefore has no historic cost of production or replacement cost. However, holdings of land do appear as a major part of the value of assets of companies in company accounts, and do appear in the Balance Sheets in national accounts, where an attempt is made to measure sectors' net worth or wealth. In the *Blue Book* Balance Sheets, 'in principle all marketable financial assets and land and buildings [are] included at market value'. Other assets for which no recognised market existed, for example, plant and machinery, were included at their 'most useful' valuation, in this case replacement cost net of depreciation (CSO, 1985, p. 202).

The implication of the treatment of land in the Blue Book is that the productive potential of the land as a factor of production is constant and therefore it does not have to be measured. However, environmental economists have recently begun to point out that the productive capacity of land and natural resources may actually be seriously depleted through time, if natural resources are used and not replaced, or if the surface of the land itself is used so intensively or so polluted as to lose permanently its fertility or usefulness. This stock of what may be called *natural capital* has a broad definition and consists of land, which is owned as property, as well as the air and the sea, which are not owned in the same sense. To economists, once a natural resource, such as water in a mineral spring, or natural gas, becomes someone's property, and capable of yielding its owner a rent, it becomes 'land', regardless of its physical form. From these simple observations flow many significant consequences.

The capital stock in the National Accounts

In the relevant chapters of the *Blue Book*, capital stock is divided into two main categories in the national accounts. The first is durable capital stock owned by the government, corporate and personal sectors. Durable capital stock consists of machinery, vehicles, dwellings and 'other buildings and works'. The second category, referred to simply as 'stocks', is held by the corporate sector only and consists of the stocks of materials, work in progress and unsold goods. These two categories correspond, respectively, to economists' concepts of fixed and circulating capital. Like all measurements of stocks of any kind, they refer to the amounts in existence at one moment – in this case, at the end of the calendar year.

For an idea of the rough relative magnitude of fixed and circulating capital stocks in the whole economy, we can compare figures in the national accounts. In 1992 total gross fixed capital stock was £2,531 billions, valued at 1990 replacement cost. The total value of stocks held was £121 billions, at current (1992) market prices. Because the value of stocks is relatively so small, most analysis of the capital stock focuses exclusively on the fixed capital component, and we shall do the same, until we come to look at the capital assets of construction firms, where stocks can loom large relative to fixed capital.

Of that gross fixed capital stock of £2,531 billions, £1,791 billion or 71 per cent consisted of built stock, including dwellings and other buildings and civil engineering works. The remaining 29 per cent consisted of all the stocks of plant and machinery (25 per cent) and vehicles (4 per cent) in use in industries throughout the economy. Unfortunately these figures contain some inconsistencies. Because of the inclusion of dwellings as part of the built stock, the figures for built and other fixed capital stock do not in one sense compare like with like. Machinery found within dwellings (kitchen equipment and so on) is not included in the stock of plant and machinery, nor are cars owned by households included in the stock of vehicles. Instead these are classed as *durable consumption goods*, on the grounds that, unlike dwellings, they have left the circuit of commodity production and exchange. They will not be used to produce saleable outputs or rented out to yield an income. Similarly, one could point out that most of the income dwellings will yield to their owners is only an imputed income. In an important sense, dwellings are also a durable consumer good. Yet dwellings are included in the calculation of capital stock. Once we start measuring imputed income for one kind of durable consumption good, why not do so for others, and impute a charge made by individuals as owners of cars to themselves as users?

Part of the confusion concerning the definition of capital stock arises from the different ways of looking at the concept of capital, depending on the particular use capital is expected to fulfil. In fact there are three different concepts of capital, which could conceivably be used to calculate the fixed capital stock. For some purposes, each is valuable as an idea, but we need to be clear when we shift from one concept of capital to another.

The first concept is that all durable goods capable of producing a flow of useful services or benefits in the future, regardless of actual cash flows, should count as capital. This concept relates capital to a *stream of utility* or satisfaction. This is the argument for counting private cars and household equipment as part of the capital stock, as well as public goods such as roads.

A second definition of capital is that it is an asset whose ownership yields an income. This distinguishes capital from mere wealth. For instance, a car would then be wealth, if owned by a household, but capital if owned by a car rental or leasing company. Money in a personal bank account is financial wealth but the same money is financial capital if it is in the bank account of a business, because the business will spend it to make profits, an income, whereas an individual will spend it on personal consumption, which will yield personal satisfaction, but not an income. Therefore this definition of capital emphasises the significance of a *stream of income* in valuing assets.

The third idea takes a narrow approach to defining the concept of capital. Capital is only that wealth which is used to produce outputs of commodities, goods or services sold to users. In this case, not only would owner-occupied dwellings not be capital, but also nor would public buildings and works, including roads, schools, hospitals, etc., because this concept of capital views capital in terms of a *stream of production of commodities*.

In spite of the problems just discussed concerning the definition of fixed capital stock, the figures given in the Blue Book show one thing of particular relevance to the study of the economics of the built environment. Namely, as mentioned earlier, over 70 per cent, of the total national stock of fixed capital is comprised of built stock. Anything which affects the value of the built stock, therefore, will clearly have a great effect upon the value of the total national capital stock, and, most probably, vice versa. Since the real capital stock is one of the factors of production, it is a source of an economy's productive capacity. The larger this stock the greater will be potential output. Economic growth is the increase in actual output over time, usually measured by the rise in GDP, and is limited by the rate of increase in capacity or potential output.

Types of fixed capital stock

Capital stock can be defined in a variety of ways depending on the use of the concept we intend, just as different tools have different applications. To begin with we need to distinguish between directly productive and indirectly productive capital stocks.

Directly productive capital stocks are those which are used by industrial capitalists to produce output that is sold as a commodity on the market. The capital stock is then directly productive of the exchange value of that output. Normally, industrial capitalist enterprises will only add to the stock of fixed capital they own, if they are confident that they will immediately be able to use it fully in production. An addition to the stock frequently indicates an addition to the use of that stock to produce an increase in output.

Indirectly productive capital stocks are those which are used to add to the productive capacity and output of other enterprises besides their owner. Thus new roads add to the productive capacity of all enterprises in a position to use them. However, they do not produce an output or an income for their owner, if the owner is in the public sector such as the Ministry of Transport. With indirectly productive investments, the public sector owner has to decide on behalf of users how much capacity they will require. The choice for government may lie between excess capacity of, for instance, the Humber Bridge, or insufficient provision, as can be seen with the congestion on the M25 motorway.

It is therefore possible that there will be a provision of indirect capital stock ahead of demand, for one of three main reasons. The state may wish to stimulate and encourage increases in private investment and output and provide new roads or universities or industrial estates as a way of doing this. We may think of this as 'you-may-take-a-horse-to-water-but-you-can't-make-it-drink' public investment, which may or may not achieve its objective. Alternatively, the state is often in a position to take a long view and may therefore commission a bridge with capacity to meet projected demand in twenty years' time, because it is technically more efficient overall to do this rather than build a smaller bridge now and then another in five years' time, and then another, and so on. A third reason for the possible excess capacity of public sector provision of indirect capital stock is because the state may be simply over-optimistic in its forecasts of public demand. The argument is that there is likely to be a bias towards over rather than under estimation, because estimates of what is likely, get confused with desirable targets in government forecasts of growth in the economy and hence in demand for these indirectly productive facilities.

It is even more likely, perhaps, though for a quite different kind of reason, that there will be an under-provision of such infrastructure. This is likely if the government does not wish to raise taxes or raise its financial deficit (PSBR). In that case, government is more attentive to financial market and general taxpayer interests, than to infrastructure user interests.

The concept of *infrastructure* is often used to cover both this kind of indirectly productive investment and capital stocks of a rather different kind, which do produce a saleable output for their owners but yet have a general influence on the capacity of the whole economy. Additional electricity generating capacity is an investment of this latter kind, for, whilst it does indeed produce a saleable output for electricity companies (and thus is directly productive), it also adds to the total productive capacity of the whole economy, in a way that would not be true of, say, a new shoe factory. Because infrastructure exists in all parts of a country to a greater or lesser extent, it becomes a major element in the decision making process of virtually all firms and individuals, when choosing to locate in a particular area. We propose to use the concept of *spatial infrastructure* to cover all capital stocks which have this kind of effect.

Like other kinds of buildings and works, but unlike the vast majority of inputs to production, there is no possibility of importing spatial infrastructure. While the electricity itself may be imported, the infrastructure by which it is conveyed, the National Grid, must be located in the country in order to reach potential consumers. The distribution of the stock of spatial infrastructure therefore places an upper limit on the potential output of a whole national or regional economy, made up of all the enterprises and industries in that national or local economic space. Spatial infrastructure is therefore to be distinguished from ordinary items of capital stock, which we shall call free-standing. Free-standing capital stock only affects the capacity of its owner's firm, because its outputs, however generally required as inputs to other industries, can be supplemented by imports.

An intermediate case between directly and indirectly productive capital consists of industries which, though they produce a saleable output, operate at a loss because their owners are able to capture, in the prices they charge, only part of the value of the benefits their investment produces. Railways, and public transport systems generally, are a classic example. The general effect is known in economics as that of *positive externalities*. For example, all road users would benefit from new metro lines, which would reduce road congestion. But road users in a city cannot be made to pay the owner of the metro for this benefit. It may be that revenue from fares alone will not cover the costs of a new line, but that total economic benefits from the line would nevertheless greatly exceed its cost. This, of course, is one of the main arguments in favour of government subsidies for public services.

Just because capital stock exists does not mean that it will be used, and obviously it is only utilised capital stock that will contribute to output. We do not actually know whether under-utilisation, taking one year with another, is more prevalent in one sector of the capital stock or another. What we do know is that it is a recurrent problem for many industries and that under-utilised capacity plays an important role in determining the rate of profit of firms. Bowles and Edwards (1993) for instance highlight the issue in manufacturing industry. The stock of buildings is also prone to over-capacity especially during recessions and the problem of under-utilisation of the built stock may even be a greater problem than the under-utilisation of plant, machinery and vehicles.

This is because of the greater planned life of buildings. If the stock of machines turns out to be in excess of what is needed, one or two years of non-replacement of machines when they are retired will be sufficient to achieve a major percentage reduction in that stock. However, at present rates of actual housing retirement, through demolition or conversion to non-housing uses, it would take about one hundred years of non-replacement of retired dwellings to eliminate 10 per cent of the housing stock. Because of the durability and the spatially fixed nature of buildings compared to machinery, the nature of the capital stock invested in the built environment is relatively inflexible in the long term. Because of the length of time needed to adjust the stock of buildings, the built environment is prone to over-capacity at certain times and in particular locations.

Machines and plant are designed to have an *engineered* capacity which sets a fixed upper limit to their capacity to produce output. A pipeline can only carry a certain maximum flow; a machine can only run at a certain maximum speed. However, *economic* capacity refers to the level of utilisation at which the cost of production per unit is at a minimum, and in many cases this may well be rather less than the engineered capacity. Beyond the least cost volume of use, costs per unit may then increase rapidly with further increases in utilisation. If a firm tries to expand its output in the short term without investing in machinery it will eventually hit the steeply upward sloping part of its cost curve and then it will hit the engineered capacity of its stock of machines. The main way the engineered capacity can be approached, and the main alternative therefore to investment, is to work the machines for more hours per day or per year by, for instance, introducing overtime or double shifts for the workers.

Buildings are rather more flexible with regard to the maximum use that can be made of them per hour. Layouts of work stations and departments can be re-arranged. Home working by staff and desk sharing can be increased. More workers and more machines can be squeezed in, albeit at a cost of lost efficiency in the use of labour and machines. Expansion of university education during the 1980s and 1990s has meant that the same university buildings, for example, are often being used to teach many more students than they contained prior to the growth in student numbers. Greater numbers use the buildings than they were designed to accommodate. This reflects the flexibility of the capacity of buildings in the short term. It is not possible to predict what increase in the floor area of the built stock of any industry would occur over a period even based on accurate forecasts of its output, without the risk of serious error.

Over the long term, the direction of technological change in many industries has been to reduce the amount of built stock in terms of floor area (though probably not in terms of value), that is required to produce the same output as before. In part this is because built space has become an increasingly expensive commodity to buy, and there is always a special incentive to re-design production methods in the long term to economise on the use of an increasingly expensive input. Thus, much of the impetus behind new hospital technologies of treatment has been the desire of the health authorities to reduce the time spent occupying hospital space by each patient. The result has been that more operations than before can be performed in fewer hospitals, and thus demand for hospital building has been reduced below what it would otherwise have been.

The valuation of fixed capital in the National Accounts

In the national accounts fixed capital is valued in two different ways. In 1990, the gross value of fixed capital at 1990 replacement cost was £2,410 billion, whereas the net value at current replacement cost was almost £1,000 billion less at £1,547 billion. The value of fixed capital therefore depends on the method of valuation and the method of valuation selected depends on the use to which the valuation will be put.

Valuation at gross replacement cost is called *gross* because it takes no account of depreciation of items of capital stock. Each item is valued as if it were new. *Replacement cost* is used to value items just as with a superior domestic household contents insurance policy, in which items are valued not at their original cost, but at what it would now cost to buy replacements. If these items are still being produced in unchanged form, this is straightforward to calculate. If not, the replacement is envisaged as an item with the equivalent productive capability per annum.

The concept of the gross capital stock is useful in measuring the productive capacity of the economy. The underlying idea is that a machine or building continues to yield the same contribution to output each year regardless of its age, until it reaches the limit of its useful life, when this contribution falls to zero and it is scrapped. If we want to measure the capacity of, say, an airline to carry passengers in the coming year, we would count up all its stock of planes. Each Boeing 747, or whatever, would have the same capacity whether brand new or ten years old.

In order to remove the effect of price changes, all items in the gross capital stock are valued at constant prices, using a common year's prices (in this case, those of 1990). The purpose of using constant prices is to measure changes in the volume of the capital stock from one year to the next. This shows the rate at which the productive capacity of the capital stock is increasing.

If, on the other hand, we wish to measure the wealth of a company, the value of its capital in the sense of what it would be worth to a buyer of shares in the firm, then we would measure fixed capital at net value, which takes depreciation into account. Older assets would be worth less than newer ones, because the stream of income to be expected over the whole of the remaining lifespan of older plant and machinery would be less.

In order for depreciation provisions to be sufficient to pay for the replacement of capital items when they are retired from the stock, the depreciation should be calculated as a percentage of replacement cost and not historic cost. Because of inflation, replacement cost will normally be much greater than historic cost. However there are cases, like computers, where the effect of rapid technical progress in increasing the output of computing power obtained from the same real amounts of resources required to make the computer, more than offsets the effect of inflation, so that replacement cost becomes lower than historic cost.

Net capital stock should be valued at current replacement cost. Inflation in the replacement cost of capital items will then increase this valuation, but that is appropriate, because the same inflation will actually increase the monetary worth of the capital stock. Net capital stock at current replacement cost shows what fixed capital stock is worth on a particular day.

Stocks of capital and flows of investment

The value of the capital stock is related to the values of investment and disinvestment flows. Investment increases the stock of capital while disinvestment diminishes it. This can be seen in (5.1), which shows gross capital stock is equal to the gross capital stock in the previous period plus gross investment (which is called 'gross fixed capital formation' in the *Blue Book*), less retirements:

$$GCS_t = GCS_{t-1} + GFCF - RCS \qquad (5.1)$$

where

GCS_t = Gross capital stock at time, t

GCS_{t-1} = Gross capital stock at time, $t-1$

$GFCF$ = Gross fixed capital formation during the period $t-1$ to t.

RCS = Retirements from the capital stock during the period $t-1$ to t.

$$NCS_t = NCS_{t-1} + GFCF - FCC \qquad (5.2)$$

where

NCS_t = Net capital stock at time, t

NCS_{t-1} = Net capital stock at time, $t-1$

FCC = Fixed capital consumption during the period $t-1$ to t.

In (5.2), net capital stock is equal to the net capital stock of the previous period plus gross fixed capital formation during the period less fixed capital consumption during the period.

Retirements measure the physical removal of obsolete items from the capital stock and therefore imply the concept of replacement investment, the amount of investment needed in a year to keep the gross capital stock, and its productive capacity, at the same level as before. GFCF in excess of retirements is sometimes called additive investment, because it adds to total productive capacity.

Capital consumption as a concept is related to depreciation, and measures the partial loss of worth of items that remain in use as part of the stock. Thus while a part of the capital stock is retired in any given year, and therefore then disappears altogether, all of the stock is subject to depreciation or capital consumption.

The change to the net capital stock is GFCF minus capital consumption and is known as net fixed capital formation (NFCF). NFCF measures the increase in the net worth of the fixed capital stock and it is the *ownership* of fixed capital to which it applies, because it is a measure of wealth. Equation 5.2 could therefore be written as:

$$NCS_t - NCS_{t-1} \ = \ GFCF - FCC \qquad\qquad (5.3)$$
$$NCS_t - NCS_{t-1} \ = \ NFCF \qquad\qquad\qquad (5.4)$$

where

$$NFCF \ = \ \text{Net fixed capital formation during the period.}$$

In 1992, according to Table 5.1, the government sector owned around 25 per cent by value of the total stock of wealth in the form of fixed capital, while the corporate sector owned around 40 per cent and the personal sector owned around 35 per cent. The personal sector's wealth is overwhelmingly in the form of dwellings, and much of it therefore actually yields only imputed incomes, whilst much of the government sector's wealth consists of public goods that yield no income at all to the government, imputed or otherwise.

Buildings and works, including dwellings, at £1,120 billion accounted for 73 per cent of the total stock of wealth. Even in the corporate sector, including manufacturing industry, all buildings and works at £278 billion were only marginally less than the value of plant and machinery at £297 billion.

If one compares the figures in Table 5.1 with those of a decade earlier, the most significant change in the period reflects the privatisation or shift of public corporations and their fixed assets from the public sector to companies in the private sector. Table 5.2 shows the percentage of all fixed assets held by public corporations, the private sector, and local and central government in 1982 and 1992 at current replacement cost. While the share of private sector holdings of fixed assets rose from 57 per cent to 75 per cent, the public sector's share of

Table 5.1   Net capital stock, by sector and type of asset at current replacement cost, 1992

Type of asset	Private sector			Public sector			Total
	Personal sector	Corporate sector		Public corps.	Government sector		
		Industrial and commercial companies	Financial institutions		Central govt.	Local authorities	
	Personal						
	£ billion	£ billion	£ billion	£ billion	£ billion	£ billion	£ billion
Vehicles, ships and aircraft	8.8	38.9	3.9	4.4	0.7	1.4	58.1
Plant and machinery	21.8	285.0	12.5	20.9	12.2	6.2	358.6
Dwellings	461.9	10.0	-	15.7	4.4	90.8	582.8
Other buildings and works	36.9	231.6	36.8	42.0	82.2	108.5	538.0
All fixed assets	529.4	565.5	53.2	83.0	99.5	206.9	1,537.5

Source: National Accounts, (1993 edn).

Table 5.2   Percentage of fixed assets, 1982 and 1992, at current replacement cost

Sector	1982 %	1992 %
Public corporations	17	5
Public sector other than public corps.	26	20
Private sector	57	75
Total	100	100

Source: National Accounts (1993 edn).

fixed assets almost halved from 43 per cent down to 25 per cent, partly because of privatisation and partly because of policy restraints on public sector fixed capital investment. As a result the public sector as a whole reduced its percentage share of demand for construction work but contractors continued to carry out building and civil engineering work in telecommunications, water, electricity, gas and other utilities, transferred to the private sector. This shift in work from the public to the private sector has implications for procurement methods in those particular construction markets.

Retirements of built stock, life-spans and replacement demand

When a firm replaces worn out fixed assets it makes a decision to invest. The cost of this replacement investment has been allowed for in the provision for depreciation of assets, and (if made) the replacement enables firms to maintain production at the same level as before. When firms expand production, the total or gross investment expenditure also includes additional plant and equipment. Gross investment less replacement investment is net investment, and it is net investment which allows for expansion. These relationships between the increase in the stock of fixed assets, gross and net investment are illustrated in the equations below:

$$I_g - I_r = I_N \tag{5.4}$$

where

I_g = gross investment
I_r = replacement investment
I_N = net investment

$$I_g = I_r + I_N \tag{5.5}$$

$$\Delta FCS = I_r + I_N - R \tag{5.6}$$

where

ΔFCS = change in fixed capital stock (usually an increase)

R = retired stock.

Substituting (5.5) into (5.6):

$$\Delta FCS = I_g - R \tag{5.7}$$

and transposing terms:

$$I_g = \Delta FCS + R \tag{5.8}$$

Thus, gross investment is equivalent to the increase in stock plus retirements. Hence, demand for building and civil engineering work is related to the existing stock of the built environment. The larger the existing built stock, the larger the annual volume of replacement demand. Assume the average life of built stock is 50 years. Then each year 2 per cent of the built environment would be retired or taken out of use, and require replacement. In this way after 50 years the whole built environment would have been replaced. Otherwise, the built stock would decline and the capital invested in it would have been consumed. Replacing 2 per cent of the built stock per annum maintains the existing quantity. It does not increase the amount of built stock. In practice, it is rare for buildings to be replaced with identical uses. When buildings are demolished, it is usual to change the use of the site or increase the plot ratio, that is the ratio of total floor space area to the area of the site.

As the number of household units increase and the economy expands, extra buildings and infrastructure are required. Therefore the annual demand for construction work is comprised of replacement investment and net investment for extra buildings and works. If in one year, following several years of no growth or decline, aggregate demand in the economy increases by only 1 per cent, then if this were reflected in extra demand for floor space, there would be a need for new building work equivalent to 1 per cent of the building stock. As a result, total demand for construction work would be the 2 per cent of building stock requiring replacement and, in addition, a further equivalent of 1 per cent of the total building stock as extra new work, an increase in demand for construction of 50 per cent in one year.

This large increase in building work in response to a relatively small increase in demand in the rest of the economy is known as the accelerator principle, and is discussed in Chapter 9. The accelerator operates in all durable capital investment markets, when an increase in consumer demand leads firms to invest in buildings and plant in order to meet that in demand.

Concluding remarks

Different concepts of capital relate to the purpose to which that capital is put. The first sees capital as providing a flow of benefits in kind over time. This concept relates to the real usefulness of buildings and built structures in providing services. The second concept of capital relates to its financial characteristics in that the ownership of capital generates returns in the form of income. The third concept of capital includes only those buildings and plant which produce commodities for sale.

The first and third concepts of capital are used to determine capacity in the economy, and this is central to understanding investment and economic growth. Capital may be directly or indirectly productive. Directly productive capital, of factories and plant, while indirectly productive capital refers to the infrastructure necessary for production to take place.

In the national accounts, capital appears both as fixed capital stock and as fixed capital formation. Some 70 per cent of the capital stock comprises buildings and built structures, whereas new built environment investment accounted (in the 1990s) for around 50 per cent of all gross fixed capital formation.

Gross fixed capital formation in all buildings and works (built environment) each year is of the order of 4–5 per cent of the value of the net built environment capital stock (valued at current replacement cost).

6
Ownership of and Investment in Built Stock and Land

Introduction

In Chapter 5 we looked at capital and its relation to the stock of the built environment. In this chapter we deal with the valuation of capital assets in the accounts of a construction sector firm, and use Tarmac plc. as an example.

We then discuss the valuation of property assets and the investment criteria used to assess building projects from the point of view of the customers of the construction industry, who are the developers. It is they, who decide which projects to select, what the risks associated with any given project may be, and whether or not these risks are worth taking. We look at the financial and strategic thinking behind decisions to build or hold property.

The chapter complements the discussion on the derivation of property values by concluding with a discussion on the derivation of land values.

The book value of assets

As we have seen, the value placed on assets is variable and is therefore to some extent at the discretion of firms themselves. A firm may want to undervalue its assets to show a high return on capital, or it may overvalue its assets to show growth in its asset base from one year to the next. There are, however, certain constraints on firms over- or under-valuing their assets.

For those public limited companies with widely diffused shareholders, the danger of undervaluing assets relative to their profit potential is that a firm specialising in takeovers of other firms will spot this undervaluation and launch an unwanted takeover bid. Such a predator could offer a price per share that would seem generous to those share-holders who believed what their company had told them about the value of its assets. The offer would be worth a premium over the asset value per share, and yet would enable the predator to acquire those assets for less than their own valuation of the assets under new management.

The danger to quoted companies of overvaluing their assets is that, though it improves the apparent strength of their balance sheet, it reduces their apparent performance in terms of their return on assets. The return on assets is one measure of a firm's profitability. More seri-ously, it exposes the firm to a forced and large sudden downward devalorisation when the pretence can no longer credibly be sustained, and this may destroy shareholders' confidence in the firm. However, during economic crises and recessions many firms prefer to use the opportunity to write down the value of their assets, because it helps them subsequently to show an early recovery of profitability, (Smith, 1992).

In company accounts fixed capital is referred to as tangible fixed assets, which are valued in yet another way, using the term, 'net book value'. Net book value simply means the value at which assets are entered in the books or accounts of the company. For instance, accord-ing to its balance sheet the tangible fixed assets of Tarmac plc at the end of 1996 were valued at £1,406 million. In notes to the accounts, the values of tangible fixed assets were broken down into three sub-headings. These were: mineral reserves valued at £752 million; land and buildings at £165 million, and plant, machinery and vehicles at £489 million.

Mineral reserves are part of what economists refer to when they use the concept of land as a factor of production. In economics, land refers to all natural non-produced resources. However, in Tarmac's and other company accounts, the value of mineral reserves is given separately from the value of land intended for building development. This is because minerals are valued by a different method to development land, and in Tarmac's case, because they are such a large part of Tarmac's total capital stock.

Land and buildings are grouped together by Tarmac. The rationale for this is based on two possibilities. Both may be intended for sale together, though not imminently. Otherwise they would comprise cir-culating capital stock, and would not be included under fixed capital. Alternatively, both are intended to be retained together and occupied by the company or leased to users as rental property. Nonetheless, this merging of the two serves to conceal some useful information when we come to the figures for depreciation. Only buildings will be depreciated, but since there is no separate figure for the value of the stock of buildings, we cannot calculate the rate at which these buildings are, on average, being depreciated. The average annual rate of depreciation of building stock would be given by the annual provision for depreciation of buildings divided by the value of the stock of buildings at the start of the year, plus or minus any adjustment for acquisition or disposal of buildings during the year. Nor, perhaps more significantly, is it possible to see the amount, if any, by which the stock of land has been calculated to have appreciated in value since the end of the previous year. However, Tarmac did provide figures for the gross book value of land and buildings at £206 million, and the gross book value of depreciable assets within that at £131 million.

Tarmac values its plant in its accounts at 'historic cost less cumulative depreciation'. Thus a machine that cost £1 million when new and is now seven years through its planned economic life of ten years might be valued at £0.3 million if accumulated depreciation were £0.7 million. The amount of provision for depreciation per year in fact depends upon the net value of the stock at the start of the year and the assumed average lifespan of the items comprising that stock. Plant is, of course, assumed to have a much shorter life than buildings.

Tarmac began 1996 with plant valued, gross of any depreciation and at historic cost, at £747 million. On this it allowed £63 million for depreciation during the year, a figure which is consistent with an assumed average life of around 12 years. However, because of inflation, annual historic cost depreciation provisions would not actually be sufficient to pay for the replacement, each year, of one-twelfth of this stock.

If we look at the accounts of a *pure* contractor, we can usefully compare the depreciation cost on plant and machinery owned by the firm with plant hire charges paid on items used but not owned by the firm. This would normally demonstrate that contractors obtained by far the greater part of the stock of plant used by hiring. Plant hiring strategy is typical of UK construction firms. Plant then disappears from the financial accounts of the contractor as an asset, owned by the contractor, but it still has a hidden presence in the production account as a factor of production. The magnitude of capital equipment used can only be measured by adding together depreciation and hire charges.

Devalorisation

Buildings are items of fixed capital stock and, like any other pieces of fixed capital, can lose their value over time in two distinct ways. One is the steady and expected process of depreciation. The other is the process we call devalorisation, which means the unexpected, often sudden, downward revaluation of fixed assets (Harvey, 1982; Smith, 1984). This occurs when there has been a downward reappraisal of the value of future profits to be expected by the owners of these assets. It therefore also means a reduction in the total book value of the capital of the firms that own these assets, and a corresponding loss of market value of the financial liabilities and equity shares of these firms. Not only do the shares in firms fall in price, the perception of the quality of loans to these firms is also revised downwards as the real assets, which provide security as collateral against borrowing, decline in value. The devalorisation of assets means that there is a greater risk of default, if assets need to be sold to repay loans. It therefore becomes more difficult and expensive for firms to obtain finance.

Devalorisation operates unevenly in different sectors of the economy, both across the stock and through time. Devalorisation can be the result of rapid economic restructuring or can be caused by an economic or political crisis. Moreover, phases of devalorisation of different types of assets are bunched into certain periods, whilst other periods are relatively free from crisis and restructuring.

Property market crises are classic examples of devalorisation. Expectations, regarding future rent levels for virtually all of a certain type of property, such as offices, are suddenly revised downwards. Confidence and expectations about future rental income affect the current valuations of property, even when no downward trend in rents has necessarily begun.

Another important cause of devalorisation involves unforeseen and rapid technological change. The opening of the Channel Tunnel, for example, caused a devalorisation of other fixed assets, such as passenger ships and port facilities on the English Channel and led to a crisis in the cross-channel ferry industry.

A third source of devalorisation, especially of buildings, concerns geographical restructuring, where competition from rival producers in other locations undermines the profitability of producers in a certain type of location. For instance, sudden economic obsolescence resulting in devalorisation of the fixed assets of existing high street retail property can be caused by out-of-town hypermarkets. Similarly, foreign textile and clothing producers can devalorise Lancashire textile mills.

Finally, general economic slumps, such as those in the UK after 1929 or 1989, can lead to a general downward revision of profit expectations for all firms, and hence a general devalorisation of assets. Devalorisation can also be a phenomenon of financial markets, in relation to the real assets which financial assets represent. Financial markets are markets in paper, where transactions concern the price and transfer of ownership of financial assets, without any change in the real assets. Devalorisation then occurs in financial markets, following speculative bubbles, which arise when confidence in short term price increases fuels the purchase of property or financial assets by raising their current prices above levels *explained* by the current profits and cash flows of the owners of those assets. This last kind of devalorisation does not depend upon a prior downturn in the real economy.

Speculative bubbles are phenomena of expectations and liquidity. In periods of great general confidence in the economic future, it is possible for promoters to market, at ever rising prices, financial assets whose value is more or less entirely dependent on the continuation of the economic boom, for they are not as yet backed by substantial current profit cash flows.

The classic examples in the literature (Kindleberger, 1978; Minsky, 1986) include shares in the South Sea Company, various railway and mineral shares, pseudo-commercial paper (for example, bills of exchange issued by merchant banks in excess of the volume required to finance actual commercial transactions) and speculative bank liabilities. Recent examples would include UK office and house property in the late 1980s.

So long as these financial liabilities retain holders' confidence, they can and do circulate as means of payment. In such booms cash itself is hardly needed by speculators to settle their obligations. The supply of credit instruments is expanding fast, credit seems 'as good as cash', and the shortage of actual cash is not felt as an impediment.

Speculators buy financial assets (or property) 'on margin', i.e. on credit. Asset prices rise. Others seek to share in the easy gains to be made, and thus rising prices and rising demand chase each other up in an upward spiral.

Then, the bubble bursts. Confidence that various kinds of financial asset are in fact 'as good as money' evaporates, as do expectations of further rises in asset prices. Holders seek to cash in their gains. The collapse may begin with a bankruptcy of a single large speculator or bank.

Suddenly banks review their lending to other speculators. A chain of demands for settlement or repayment in cash begins, and thus a scramble to sell financial assets, at any price, to obtain the necessary cash. But these assets find no cash buyers. Suddenly everyone wants liquidity, and speculative assets transpire to be illiquid, for they have no buyers. There is a wave of bankruptcies, bad debts, asset write downs and devalorisations.

As well as occurring in financial markets, devalorisation of capital assets can also take place in real markets, which are concerned directly with the production and sale of goods and services. When there is a larger total stock of fixed capital in an industry than can be fully utilised, one of two things will happen. All firms and all production units can share out the excess capacity between them, more or less equally. Alternatively, competition between firms and rationalisation of the allocation of output to units within a firm will result in some firms and units being closed completely. If firms all work at less than full capacity, then the profit expectations of all firms will fall in about the same proportion. If, on the other hand, firms decide to compete so that only the survivors can operate at or near full capacity, then some assets will lose all their value, except as scrap, whilst others will keep most of theirs. The surviving assets will not fully maintain their previous valuations, because in order to force closure of the least profitable competitors, it may have been necessary to reduce the general level of prices and profit margins (Brenner, 1998).

Schumpeterian innovation, super-profits and revalorisation

Whilst innovation can force devalorisation in the assets it makes obsolete, innovators can themselves expect to make above average profits, until their innovation is widely imitated or improved upon. The capital assets of a firm making super normal profits due to an innovative lead it has over its competitors, may be revalued upward in terms of their net book value, and will also be revalued in the stock market, to a level far higher than their historic cost. The key point is that an innovator's fixed capital assets will immediately have a value that bears no relation to their cost of production or purchase price.

For most goods, competition ensures that the selling price of a good bears a fairly close relation to its cost of production plus a mark up for the normal profit of the producer. This is normally the case too for capital goods purchased by innovators, as plant and machinery suppliers compete with each other to provide these capital goods.

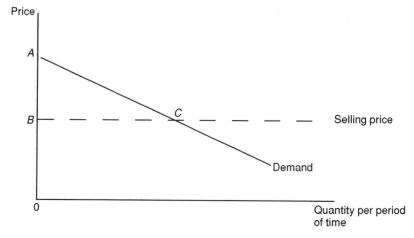

Figure 6.1 Diagram showing consumers' surplus, *ABC*

However, it is not true for the price of goods produced by innovators. It is possible to explain this is the terms of the neo-classical approach of a static equilibrium world. We assume in Figure 6.1, that the value of any good to its marginal or least keen purchaser just equals its selling price. At the same time other purchasers obtain some consumers' surplus, the triangle *ABC*, which is the excess of their valuation of the good over the price they have to pay for it.

In Figure 6.2, assuming all firms are profit maximisers and incur the same costs, the marginal revenue from the last unit of output, n_1, from each producer just recoups its marginal cost, while pre-marginal units yield a surplus of revenue over their cost, represented by area *DEF*. This surplus is equivalent to consumers' surplus in the sense that firms would be willing and able to pay a higher price than the market requires for the inputs required to produce this pre-marginal output. One way of looking at innovation is to say that the innovator obtains an exceptionally large producer's surplus, area *DGH*, from the inputs they purchase, because innovation shifts costs downwards towards MC_1, and the innovator increases output to n_2. The innovative producer's surplus increases by the area *EGHF*. Eventually, as innovations are imitated or copied and become widespread, competition drives output prices down and the marginal revenue curve shifts to MR_1. Consequently, the innovator is forced to reduce output to n_3. At the same time competition to buy the required inputs drives their prices

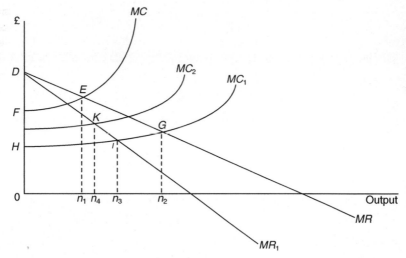

Figure 6.2   Diagram showing simplified producers' surplus and the effects on costs of product and process innovation

up, and the marginal cost curve shifts once more, this time to MC_2 and output moves to n_4, at the intersection of MC_2 and MR_1 at K.

Process innovation enables firms to lower their costs from MC to MC_1, below those of their competitors, without at first affecting their MR curve. This enables the innovator to capture the area of $EGHF$ in Figure 6.2 in the form of additional profit. This temporarily increases the producer's surplus and the firm's profits until competitors adopt the new process and price competition reduces profit margins. In the meantime, the valuation of the assets of any firm with increased profits will be raised to reflect the expected duration and extent of these super-normal profits. However, innovation in the production of the built environment sector is relatively slow. Consequently, innovators' super profits are relatively rare for construction firms or for their clients (Bowley, 1966; Salter, 1969).

Techniques of valuation of buildings

In 1991, the surveying profession was given the task of evaluating the housing stock in the UK for the purpose of assessing the amount of property related tax due in Council Tax. The values which buildings were given were based on assumed market prices in 1991. Market values are exchange values.

However, Bon (1989) points out that economic goods have both an exchange value and a use value to their owners. The use value defined here as willingness to pay is the value placed on consuming the good by the purchaser. This distinction between exchange and use values, he argues, also applies to buildings. Indeed, property speculation is only possible on the assumption that sooner or later an occupier can be found, who is willing to pay to use any given building. Property speculation is based on market or exchange values being less than the use values to potential occupants. It is possible for exchange values to be greater than use values, but this results in empty property or reductions in price.

Property valuation techniques are used by surveyors and developers to establish budget requirements for projects, for asset valuation in company accounts and for tax purposes. Valuing property assets allows comparisons to be made with alternative financial assets in a company's portfolio.

Valuations are either market tested or expert. Shares and most other financial assets have the advantage, from the point of view of valuing them, that there is a daily, actual market in which transactions in these assets occur. Each share in a company is homogeneous with all other shares of that company, and can reasonably be assumed to command the same price. Market prices therefore exist, and current market price, though liable to fluctuate somewhat from day to day, can be known. Such assets are usually valued in their owner's books at market value or acquisition price, whichever is the lower, and this has a clear practical measurement. Problems only arise where the amount of shares to be valued is large relative to the total volume of transactions in that share. In this case, if a large bundle of shares were suddenly to come onto the market it could depress the market price.

Expert valuers, on the other hand, attempt to estimate what the market value would be if an asset were to be sold at the date of the valuation. An estimate is all that is possible, with the risk of over or under valuation, because items of real estate property always have characteristics of location, design and state of repair, which make them unique. Therefore, although some property transactions occur each day, it is by no means the case that the prices of those transactions can be assumed to apply to any property in question.

Payback period

Payback period is the number of years it takes for extra annual income resulting from the investment to equal the investment cost of an item of real capital. Purchasers of financial assets, like shares, will not usually think in terms of payback periods – how many years it takes before the income from the share equals the outlay on purchasing it – because in this case the resale value of the financial asset will also be important. Many corporations use a payback period in their invest-ment decisions, where it can be appropriate when the value of the asset is expected to decline rapidly through time, so that both resale value and expected long term profits from the asset are low. This is not normally the case with building projects, though it may apply to some construction equipment.

Related to the idea of a payback period used by a decision maker is the idea of an investor's *time horizon* – the idea is that one only takes account of those things that are within that horizon. The length of time horizon used will depend upon the lifespan the individual or institutional investor believes it will have, and also, perhaps, on some sense that the problems of the near future are so urgent that the more distant future will have to be 'left to look after itself'.

Short time horizons work against investment in long-lived buildings and works, since many of the benefits are completely disregarded on the grounds that they lie beyond the relevant time horizon. At the same time, it has proved hard in many cases to find significant current cost savings by deliberately designing buildings for a shorter life. In any case, many building and planning regulations are applied *as if* all buildings were supposed to be long lived.

Life cycle investment appraisal models

Life cycle models suppose that investors making decisions to build take into account the costs and benefits of construction projects throughout their useful existence, including their disposal or demolition. Thus, as well as estimates of construction cost and the cost of borrowing to pay for construction, estimates of the costs of maintaining and operating the built structure are taken into account. Estimates of additional revenue resulting from the project are extended to cover the whole life of the building, possibly including revenues and costs from rebuilding on the same site.

Like Keynes, we argue that all investment is largely 'irrational', because no investor can really 'know', even approximately, what the revenues or operating costs of a project will be even five years into the future. Given the extent of uncertainty that exists about the future, it can be in effect irrational (a waste of time and nervous effort) to attempt too much 'rational' forecasting of the very uncertain distant future. Attempts to date at very long range forecasting of prices and costs have more often than not performed no better than would a simple assumption that 'the future will be like the present' (for example, both 1950s and mid-1970s forecasts of 1990s energy prices).

Another reason investors tend to ignore the more distant future of their projects is that they choose or are forced to apply high 'discount rates' in their decision making. The present value of a £1 million cost or benefit occurring in 20 years' time, with a discount rate of 10 per cent is only £163,508, and at 15 per cent is £70,265. Nevertheless, life-cycle economic models of project proposals can be of use in bringing together the large number of variables involved in any project, ensuring that at least some of the consequences and implications of decisions to build are considered systematically (Gruneberg and Weight, 1990).

Discounting

Discounting involves taking the time value of money into account by recognising that money payments in the future are worth less than money payments made in the current period. For an introduction to the techniques involved in discounting, see Gruneberg (1997). What determines the discount rate that an investor will use in appraising a construction project? The current level of 'real' (i.e. over-and-above current inflation) long-term interest rates is one factor. More generally, we can think either of the 'cost of raising capital' to the investor (interest on loans, dividends on issues of share capital) or of the 'opportunity cost' of tying-up capital in a project, and therefore having to forego the 'opportunity' of other investment schemes.

In boom periods, which often coincide with monetary regimes of easy credit, a large supply of loanable funds, and therefore relatively low real interest rates, the firm's estimate of its opportunity cost of capital is likely to be higher than the cost of raising capital, because in booms firms believe they have many highly profitable possible uses for capital. The exception will be firms prepared to borrow without limit, for whom all projects expected to yield more than the rate of interest can be undertaken simultaneously.

In the last part of booms and the early part of recession periods, on the other hand, real interest rates may be high – deliberately pushed-up by central banks to reduce the inflation provoked by the preceding boom. There will seem to be few good investment opportunities, and so the opportunity cost of using finance may fall and the interest rate becomes the crucial deterrent to investment.

Different investors use different discount rates. From time to time the Treasury lays down the discount rate to be used by the government sector to appraise its investment proposals – e.g. for road or hospital building. There has been an extensive academic debate around the question of the proper criteria for determining the government sector discount rate. The argument for a rate set lower than the long-term real market interest rate revolves around the following points.

First, there is the idea that government investment should consider future generations equally with the present one, because government is the only entity which can represent the interests of those as yet unborn. Low discount rates will encourage investment in general and, more particularly, in projects whose benefits lie in the distant future. This will be at the expense of a lower level of current consumption – resources will be diverted from current consumption to investment intended to increase consumption in the distant future. Governments may regard it as their task to correct a bias in private investment towards the presently-living and towards the short rather than the long term.

Second, the state (though not particular governments) may regard itself as having an indefinitely long existence. It may therefore be prepared to build for posterity, in a way that not even the most established of corporations will.

Third, market interest rates are as high as they are in part because they contain an element to compensate the lender for the risk that the borrower will default on the loan. When stable governments of advanced capitalist economies are the borrowers, this risk hardly exists. On the other side are those who argue that, if government is allowed to cost its borrowing at a lower discount rate, this will cause scarce investible funds to be diverted from more financially viable private schemes to less financially viable public projects.

Private corporations normally use discount rates substantially higher than the long-term real rate of interest. If the firm operates in a sector where it is protected from competition, where demand for their product is more or less guaranteed to continue and the rate of technological change or innovation is low (like, for instance, water companies), then they should, and probably will, use a lower discount rate and a longer time horizon than firms in fast-changing, competitive and unstable industries.

This general use of higher discount rates by firms than governments appears to reflect a shorter-term time horizon among both their managers and their shareholders. Few of either are prepared to wait for twenty years or more to see the benefit of an investment in the form of higher profits and dividends. Nor do stock markets seem to value as highly shares whose profits lie in the promised distant future as they do shares with higher short-term profit prospects – in other words, stock markets reflect a collective strong time-preference for the near over the distant future amongst shareholders.

The creation of money and credit by financial institutions

We have shown that most purchases of new and existing built structures are financed by the purchaser's borrowing from financial institutions. Financial institutions are divided into banks and others. Put simply, banks are able to create credit. That is, they can lend-out greater sums than are deposited with them by savers. Non-bank financial institutions, like insurance companies and pension funds, essentially act as simple intermediaries between savers and borrowers. They are only able to lend as much as savers have chosen to deposit with them.

The nature of bank created credit, which Harvey (1982) calls 'fictitious capital', is that purchasing power is created, in the hands of those who obtain the credit, ahead of any production. Nor is there any corresponding reduction in purchasing power by others, as there is in the case of savings loaned-onward. It thus breaks the 'normal' identity between output and expenditure.

Because new buildings and works have to be purchased 'now' whilst their contribution to national output will come much later, and be spread over a long 'life', credit is crucial to giving buildings a current market demand and value. However, if the expected future profits or rents from the property do not materialise, then the assets that take the form of credit-loans to property owners can be 'exposed' as fictitious capital, since the capital has been created on no stronger basis than a promise of future profits, rather than out of a saving and re-investment of profits already made.

Landowners and capitalists in the development process

Having discussed the valuation of property or real estate from the point of view of developers, it is now appropriate to consider the role of land and landowners in the development process and that of labour.

Without the intervention of workers, materials and machines would not produce anything. Sites would lie idle. However, when labour is employed and production occurs, the value of what is produced is greater than the amount paid to labour and the difference between what labour is paid and what it produces is what Marx called exploitation.

To minimise exploitation, labour seeks an increased share of sales revenues, and management, representing capitalist employers, attempts to keep those wage pressures under control. If workers succeeded in eliminating exploitation, then wages would rise and prices fall until prices were just sufficient to cover wages and other costs, including the cost of materials and of depreciation or replacement of fixed capital goods used up in production. (The selling prices, and hence costs of use, of those materials and fixed capital goods would in turn just cover wage and other costs, with no margin for profit).

Profit is not a cost, but a surplus (of revenue over cost), and can be *squeezed* between upward pressure on wages and downward pressure on revenue. However, profit in any one sector cannot be squeezed much below the average or normal rate obtaining elsewhere, because money capital is **mobile**, and will simply switch from the lower- to the higher-profit sectors. This will reduce the intensity of competition between capitals in the sectors losing capital and increase it in the sectors into which capital flows, thus tending to equalise rates of profit between sectors. This mobility is crucial both to the survival of individual capital fortunes in a context of continuous economic change, and also to the 'bargaining power' of capital as a whole relative to well-organised labour – if labour threatens to push wages too high, capital can plausibly threaten to quit that sector or place.

We need here to distinguish between land and other forms of capital. The non-reproducibility of the natural resources of land and minerals makes them a special category in contrast to capital which consists of reproducible real assets. If the aim is to understand the production process, then land and its mineral resources exist separately and apart from the economic analysis of capital. Moreover, the intrinsic immobility of land is fundamental in the different determination of profits and rents.

It was David Ricardo (1973), discussing early nineteenth century agriculture, who first clearly described the process whereby labour produced surpluses part of which the capitalist tenant farmers then had to pay to the landowners in the form of rent. The more fertile the land, the greater the surpluses produced and the greater the rents paid to the landlords, leaving the rate of profit on capital the same on land of differing fertility.

The reasoning behind this set of relationships is based on the social and legal relationships of the participants. The landowners had title to the land, and in order to farm it, the tenant farmers had to compete with others in an auction. The highest bidder won the right to farm. Competition between capitalist farmers for access to land would tend to force land rents up to a level just consistent with a 'normal' rate of profit for the farmers (i.e. a similar rate to that obtained by capital in the rest of the economy). The tenant farmers then employed labourers who also competed for work. In this 'auction' for work, it was not the highest but the lowest bidders who gained employment. This competition between workers for jobs tended to keep wages consistent with those in the rest of the economy. Unemployment, more exactly the existence of a 'surplus' reserve of unemployed labour, in this view, is the main force heightening the intensity with which workers compete for jobs and thus keeping wages down. Labour mobility also plays a part, as higher wages in one sector or place may attract labour from elsewhere, thus increasing competition for jobs in this sector.

Landowners by contrast are only in a very limited sense in competition with one another. Each owns a unique piece of territory, for which other land is an imperfect substitute. Moreover, high land prices and land rents cannot attract an inflow of additional land (cf. capital and labour). Land is immobile, and cannot be produced or increased in supply. Greater revenue from production, or lower wage costs, will simply increase the competitive bid prices capitalists offer for land, and thus drive up land rents and prices, without setting in motion any compensatory or corrective force that would eventually reduce them. If capitalists react to high land prices by trying to economise on the quantity of land used, i.e. by using some land more intensively, this will simply drive up the price of that land still further, though it may certainly reduce the price of poorer (more distant, less fertile) land.

Ricardo's theory was based on an agricultural economy and the relationship between farm labourers, tenant farmers and landowners. The relationship between construction labour, contractors and other construction capitalists, and real estate owners is similar but not quite the same. The key points are: (1) that land owners as well as capitalists share in the surplus of the value of built environment produced over and above its cost of production; and (2) that competition between capitals in the construction sector does not drive the prices of property down, as would be the case with 'normal' industries – instead it merely drives up land rents whilst property prices remain unaltered. This is one of the main explanations for the tendency of property prices to rise over time relative to prices of commodities in general.

Note, by the way, that in Ricardo's theory (and in all *orthodox* economics to this day) it is not the case that high land prices 'cause' high property prices. Rather, it is the institution of private property in land, and high property prices, together with low construction prices, that 'cause' high land prices. Normally, prices are set on a cost-plus basis. When the price, and therefore the cost, of inputs used in an industry rise, this will be passed on as an increase in the prices charged for outputs of that industry. It might be thought, therefore, that land is an *input* to the production of the built environment, and that increased land prices, appearing as an increased cost to a developer, would be passed-on to the purchasers of property as higher selling prices.

However, this is not how things work, in Ricardo's theory. Most simply, this is because land rent is not a cost-of-production but instead is the landowner's share in the production surplus. If the landowner did not receive rent,[1] in the first instance property prices would be unchanged but construction capitalists would obtain greatly increased profits, as all the surplus would accrue to them. Now, a process of capitalist competition would start to bring this profit rate down towards the norm, and this would cause property prices to fall. If land were 'free' to capitalists, the built environment market would be flooded by capitalist building developers prepared to sell buildings at a price that just covered costs of production plus an average rate of profit. We call such a price the 'price of production'. Eventually this flood of cheap new buildings would tend to drive down the price of existing property. Differences in the attractiveness of, and therefore in the demand for buildings in, different locations would, in the absence of land rent, be reflected solely in differing densities of development. The higher unit cost of building to higher density would provide the limiting force distributing some demand to less attractive locations.

Marx (1970 edition) divided land rent into differential and absolute. The former refers to the premium rent that more attractive (better located; more fertile) land can command over less attractive. *Differential rent* I is based on natural differences in quality of land. *Differential rent* II is based on man made improvements, such as the effects of reclamation or of infrastructure. Investments in public infrastructure, especially in transport, can reduce rent differentials by increasing the relative attractiveness of previously inaccessible locations, or can increase differentials if the transport network is concentrated upon existing centres of high demand and density. *Absolute rent* is based on the collective monopoly power of the small land owning class and refers to the general or average level of land rent.

Rent, or the price of land, relates to scarcity of supply relative to demand. As the area or supply of land is fixed, and will not change as its price changes, its price (both differential and absolute) is determined by demand – by the willingness-to-pay of the highest single bidder for each plot or tract. However, the power of landowners in general can hold some of the land in existence 'off the market' – i.e. make it unavailable for sale or rental. In this way the supply can be restricted below its 'naturally given' level, and absolute rent on the land that is still available can be increased. 'Green belt' and similar restrictions on the supply of non-agricultural land have a similar effect.

Rent in everyday speech is the annual amount paid for use of a facility (farm, office, house). It is important not to confuse rent in this sense with Ricardo's rent, also known as *ground rent*.

Total rent paid is building rent plus ground rent. *Building rent* is simply an annualised form or version of a building's 'price of production'.

While in Ricardo's time landowners were a clearly distinct social class from capitalists (the aristocracy, church and crown owned most land), today the social distinction is less clear, with much land owned by capitalist firms, either property companies or owner-occupiers. The economic distinction between land rent and capital profit survives this social change, however. Capitalist landowners auction their land to the highest bidder in order to maximise their rent income in just the same way as traditional landowners. Indeed if anything they are likely to be more single minded and more rational in their pursuit of maximum rent, since they are less interested in non-monetary benefits of land use and ownership. However, the increasing dominance of the supply side of the land market by capitalist landowners does explain the shift from leasing or renting-out land, whilst retaining ownership, to outright sale of freeholds. Traditional landowners' sense of status was grounded in retention of ownership of *the land*, whilst capitalist firms regard it as one possible asset to hold, to be compared with others for its returns (Massey and Catalano, 1978).

Concluding remarks

We began the chapter by looking at how firms assess their asset values. These asset values reflect the accounting decisions and motives of firms at least in the short run. Moreover, the actions of others may have adverse affects on the valuation of a firm's assets if innovation by other firms renders existing plant and property obsolete. The institution of private property in land means that it is the land or property owners, collectively, who stand to appropriate much of the economic benefit that would result from more efficient design and construction of buildings. Unless they themselves become property developers, competition between construction firms is likely both to exclude them from obtaining much of the benefits to be had from innovative construction (lower cost or more valuable buildings) and, thereby, also to reduce their *incentive* to search for such innovations.

Both buildings and land acquire market values and form the assets of property companies. However, the nature of buildings as assets can be seen as a function of the expected use to which they will be put until they are demolished. Buildings therefore have an expected life cycle, beginning with the cost of construction followed by a period of use which, though it may be extended and may be uncertain, is finite. This finite existence of a building is in contrast to land which has no cost of production and will continue to exist indefinitely.

Note

1. That is, if we imagine that private property in *land* has ceased to exist, but that private property in *buildings* remains.

Part III

The Nature of the Construction Process

7
Actors and Roles

Introduction

The nature of the construction production process can be characterised as temporary in that temporary organisations are set up to build projects. Teams of professionals and contractors are assembled often at relatively short notice. Once a project has been completed, the architects, quantity surveyors, engineers, contractors and site labourers move on to other projects on other sites, where they will usually work with different sets of people and firms. In this respect there are similarities between the construction sector and, say, the film industry. However, although the construction process in a given location may be of a temporary nature, construction production itself is a permanent process, continually employed in producing and maintaining the built environment.

In this chapter we examine the characteristics of projects and the roles of participants, emphasising the temporary nature of project teams. We discuss the roles of all the participants including those of developer, designer, builder, owner and user. We conclude this chapter with an analysis of the work of contractors and their management of risk by using a portfolio of projects.

The nature of construction as a process

The construction industry is a large and important economic sector of the economy. As an industry it has several distinct economic features. Yet most of the individual characteristics of the construction industry can be found in at least one other sector of the economy. For instance, in common with agriculture, the timing and progress of work in the construction industry is dependent on climatic conditions, and like the film industry, the construction industry moves from one temporary project to the next. It is, however, the particular combination of characteristics which makes the construction process unique.

The starting point for an understanding of the economic theory of the construction and built environment design industries is that they are orientated towards individual *projects*, whereas the economic theory of manufacturing focuses on mass produced *products*. Construction, with the important exception of housing, is largely concerned with one-off production. Assembly takes place on site and as each building project is completed production moves on to new locations. Each construction project is discrete and temporary. This is the major characteristic of the construction industry and is in contrast to the greater part of the manufacturing and service sectors, which are concerned with the problems of continuous production of identical or similar standard units of output in permanent premises, such as factories, offices and shops.

It is thus customary to stress the fact that construction projects are often unique and discrete in terms of design, use and location, and that this makes production one-off or small-batch rather than mass or repetitive. But, even more important than their heterogeneity and separateness is the fact that each project is an exercise in the expression of power. The dominant actors in a project will successfully impose their definition of what constitutes the project upon the other participants – in effect, they will assert their definition of reality.

The matrix in Table 7.1 shows the variety of project types in the construction industry and the functions of the participants. Each project, (A to P), acts as a focus for the participants involved and entails a unique combination of design, production, ownership and use. Different firms and organisations and different techniques are required for different types of work, such as building and civil engineering. While the four roles or functions shown are always present, in all projects, the identity and economic character of the actors performing these roles will differ systematically between project types.

The nature of team-working in construction

What is a *project* and who has power and control over it? The term *project* refers in its general-language use to a sustained implementation of an individual act of will to undertake an activity and achieve a goal, as in, 'my project is to climb Mount Everest'. The project begins with the intention, and persists until achieved or abandoned. It embraces,

Table 7.1 Project types and functions of participants in the property development process

Functions of participants	Project types															
	Building				Civil Engineering				Refurbishment				Repair and maintenance			
	A	B	C	D	E	F	G	H	I	J	K	L	M	N	O	P
Design																
Production																
Ownership																
Use																

from the individual's point of view, all kinds of instrumental actions intended ultimately to contribute to the realisation or achievement of the project (the goal).

Individuals can truly share projects only if they share a common goal and will. For example, members of a political party might share the project of making theirs the party of government. Economic life consists of the intersection of personal and group projects. Person A's project may be to climb Everest, B's may be to become rich, and C's to survive and feed himself or herself. B may find ways of achieving his or her project by employing C to produce some good or service which B can sell to A for a profit. At the same time, by obtaining paid employment C has managed to pursue his or her own project.

Now, these different projects have different boundaries in time and space. It is possible to see A's project clearly visible, involving in this case the extraordinary movement within a limited duration, of climbing Everest, but our view of B's and C's projects may be thereby obscured.

Without necessarily realising it, we have become accustomed to seeing the construction world or construction activity as the set of projects of one particular type of economic role player, namely the developer. Developers' projects are plans to bring into being particular buildings that will be owned by them, and which they can then either sell or use. Not even the designer has the developer's power to impose a project on others. Designers' projects are to have their plans and elevations erected in reality, designs which first existed in their minds and in their drawings. Designers and developers are by no means equal in power however, for the designer is powerless to achieve a project other than on terms laid down by a developer. For example, if designers want their designs to be built they must make sure they design only for the space owned or ownable by one developer.

It is in the developer's sense that the term project has been taken and given an apparently natural and even objective meaning in the literature of construction. This becomes clearest in the language of project management, where all other participants in the construction process are assumed to sublimate their own projects and goals in the coalition of firms formed to carry out the developer's scheme. The goals of the project coalition are supposed to be those of the developer. Therefore, the goals of the project team involve completion of the development on terms that the developer judges to be successful.

Suppose, for example, that design of a developer's project is fragmented between a structural engineer, an architect and an interior

designer. The interior designers may then have as a project, the construction of an interior which successfully realises their intentions and
artistic goals. But they are concerned with the rest of the developer's
project only insofar as its achievement impinges upon the accomplish-
ment of their project. In reality, not only are the design and construction of a building normally each fragmented, but each building and
design firm will be working on parts of many developers' projects
simultaneously, and may believe that their goals are best achieved
through the overall effects of their work on that ensemble or portfolio
of projects.

Insofar as all the parties to the process simply have as their project
the making of profits, then they will take on the developer's project as
if it were their own, only so long as this is the best way for them to
achieve those profits. Firms of all kinds (contractors, architects, etc.)
will normally have as their project the overall financial success of their
firm, though they may be employing individuals whose personal projects are partly artistic, technical or professional. The developer's power
normally derives from their ownership of land and money, which they
use to convert into ownership of the finished building, though it is
true that developers sometimes only control, and do not own, the land
and money with which they develop. Because of the dominant role of
the developer, in the rest of this chapter, we will speak of 'the development process', rather than 'the construction process'.

In Table 7.2, the roles of developer, designer and builder are treated as
if they are universal and apply to the production of the built environment in all economies. However, the identity of the actors in the process
is socially determined, and varies over time and place. Thus an architect
is an actor, and architects are a category of participants, whose role is to
design buildings. However, the building design role in any one project,
and even more so in any society, may be shared between several actors
from different categories of participant. Moreover, the social meaning of
the category *architect* will change as society changes, for instance, from
'architect' meaning an autonomous individual to 'architect' being used
today to refer to a corporate entity. Powerful sets of participants in the
development process will struggle to assert the identity of actor and role,
just as architects at times have claimed monopoly over the role of building design. The state is however, the ultimate repository of power in a
society, and rules or decides on all such claims, using legal devices to
demarcate roles and functions.

We shall use this division between actors and roles in order to distinguish between the different *structures* of actors and their roles,

Table 7.2   The roles of participants in the process of providing the built environment

Actors, participants or transactors	Roles		
	Developers	Designers	Builders
DLOs[1]			
Speculative builders			
Contractors			
Construction managers			
Building workers			
Project managers			
Quantity surveyors			
Architects			
Engineers			
Property companies			
Owner-occupiers			

Note:
[1] DLO: Direct Labour Organisations, in-house construction departments used mainly by local authorities and large property-owning organisations, such as health authorities.

which co-exist in the contemporary UK. This distinction will also form a framework for studying recent change in the dominant structure of participants in the UK construction industry. In addition, it can be used in order to compare and contrast the organisational structure of the development process in different countries (Winch and Campagnac, 1994).

The persistence through time of the social meaning and status of a category of actor or participant depends upon the exercise of power and is defined by their relationships with other social actors, above all their economic relationships. Many but by no means all of these actors in the development process are firms. In order to participate in any given project, firms, government departments, organisations and individuals undertake at least part of one of the roles shown. Different firms become involved through contractual arrangements. These take the form of business transactions with at least one other party to the process. It is conventional to categorise these transactions as main contracts when they are transactions with the developer directly, or as subcontracts, when they are with some other party.

The transactors shown in the three matrices below are *examples* of firms or organisations which participate in the production, ownership and consumption of buildings and the built environment. The arrangement of firms is therefore specific in time and place to each project.

Note that, between versions of the matrix, the roles remain unchanged, whilst both the types and appropriation of roles of the actors varies.

Of course, there are many more participants involved in the production process than are included in Table 7.3a, which only shows those core roles that relate directly to the role of the developer. Moreover, the roles in Table 7.3a are those basic roles which are always and necessarily present in any development process, and give power to the actors concerned. Table 7.3b is a matrix of the same actors on the horizontal and vertical axes. Reading down each column each cell represents the type of transaction, which takes place between any two actors, and the nature of each participant's role in it, either as buyer or as seller. The row of actors supplies or provides services for the actors heading columns of the matrix.

Activities (transactions) in Table 7.3b matrix are described from the point of view of the *purchaser*, in each case (the actor at the head of the column). From the perspective of the *seller* (the actor at the head of the row), each 'purchase' of course, would appear as a sale by them. Note that, because the matrix is drawn in terms of purchases, the columns of the contractors and professional service firms are largely empty, as neither buys anything from the other actors shown here.

All developers' projects are either speculative or non-speculative. From the point of view of a developer, a project is speculative whenever production precedes sale to a building owner. Otherwise the project is non-speculative. However, from the point of view of firms in the construction industry, projects are either speculative or carried out to contract. If projects are initiated and developed by construction firms, the construction industry calls them *speculative projects*, and the developer is called a *speculative builder*, to signify that a construction firm is acting as a speculative developer as well, presumably, as taking charge of the construction of the project. If others outside the construction industry initiate and develop projects, then the building work is carried out to contract. From a construction industry point of view, it makes no difference whether the outside developer is a speculator or an owner-occupier. The construction firms are only involved as contractors and the *contracting system* applies.

This would be reasonably clear except for one unfortunate peculiarity in the way the boundary line is drawn in the official statistics between the construction industry on the one hand and the property industry on the other. The anomaly is this: all speculative developers of housing are considered to belong to the construction industry, and are described as speculative housebuilders, regardless of whether or not

Table 7.3a Actors and roles in a private new housing project, suburban England, 1980s

Roles	Actors				
	(1) Speculative builder	(2) Specialist contractors	(3) Former landowner	(4) Professional service firms	(5) Households
	Volume house-builder, probably part of a DCG.[1]	Single trade contractors	Agricultural or industrial landowner	Architect, structural and civil engineers	
Developer	Buys land ahead from (3). Obtains full planning permit for a scheme design. Develops site in parcels, over several years		Obtains outline planning permit. Sells freehold of land to (1)		
Designer	Chooses mix of dwellings and sizes from own set of standard types			Designs site layout, carries out technical specification for (1)	
Builder	Project manages and pays the specialist ('trade') contractors. Buys some materials and components	Work to contract for (1), either on supply-and-fix or labour-only contract. Employ building workers. Buy some materials			

Table 7.3a Continued

Roles	Actors				
	(1) Speculative builder	(2) Specialist contractors	(3) Former landowner	(4) Professional service firms	(5) Households
	Volume house-builder, probably part of a DCG[1]	Single trade contractors	Agricultural or industrial landowner	Architect, structural and civil engineers	
Owner					Purchases house on completion from (1) and maintains the property
User					Consumes the property by dwelling in it

Note:
1 Diversified construction group.

157

Table 7.3b Actors and transactions in a private new housing project, suburban England, 1980s

	(1) Speculative builder	(2) Specialist contractors	(3) Former landowner	(4) Professional service firms	(5) Households
(1) Speculative builder					Purchases houses on completion
(2) Specialist contractors	Employs trade firms	Employ subcontractors			
(3) Former landowner	Purchases land				
(4) Professional service firms	Engages designers, obtains cost advice and management services, employs selling agents		Engages advisers, to help obtain other planning permission		
(5) Households					

they actually perform the role of builder or contract it out. Whereas, a speculative developer of any kind of project other than housing is considered to belong to the property industry, even if they take a very active part in the building role. The reason for the anomaly is purely historical. The earliest big, modern speculative housing developers were in fact builders, like Laing and Wimpey in the inter-war period, who as construction firms tended to build the housing projects they developed. On the other hand, the earliest big, modern developers of offices were property owning companies, in the sense of being firms whose primary activity was to own stocks of property and collect streams of rents

Thus, a project may be speculative from the property developer's point of view in that its developer intends to sell it to another owner but has not sold it before starting construction. However, from the builder's point of view the work is not carried out 'on their own account' but on a contract working for a developer. Thus, a project can come within the scope of *the contracting system*, rather than the *speculative system*, even though its developer is acting speculatively – so long as the developer appoints a contractor to be responsible for construction.

The essence of the contracting system is that it involves separation of responsibilities for the roles of developer, designer and builder between three separate actors, called respectively the client, the architect and the contractor. The alternative system is sometimes called 'the speculative system' because, historically it was speculative builders who first developed it, but more accurately it should be called 'the integrated system'. It is an integrated system of building development because, in it, one actor combines two or more of these roles, as, for instance, a developer-designer or a developer-builder.

In the fully integrated system, the roles of developer, builder, designer and building owner are all subsumed within one organisation or actor. It corresponds to what economists know as *vertical integration*, where a firm expands the scope of its activities by taking on the roles of those firms selling to it or buying from it. The main integrated actors in the building process have historically been public authorities. At one time it was commonplace for a UK local authority to combine all these roles, with its own architects' and engineers' departments and its own building department or direct labour organisation. There are also some private sector examples of fully integrated systems though these are unusual in construction projects apart from speculative housebuilding. Occasionally firms like Wimpey may retain ownership as rental investments of offices which they developed and built, to

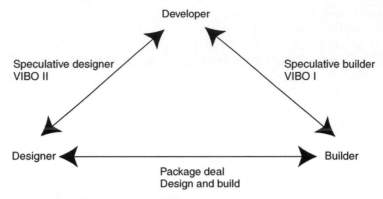

Figure 7.1   Partially integrated construction systems
Notes:
VIBO I – Vertically integrated building owner version i – developer/builder
VIBO II – Vertically integrated building owner version ii – developer/designer

become developer-builder-owners. Private Finance Initiative (PFI) projects may provide examples of high integration with the same firm taking the roles of owner, operator, designer and builder.

Figure 7.1 gives examples of partially integrated production systems, which are far more common in construction than fully integrated systems. The triangle shows the separation of the three roles while the combination of any two is given by the side connecting them. If the roles of developer and builder are combined, we can call this the 'speculative builder' version of the integrated system, provided the developer is in fact operating speculatively, and not developing the project for their own occupation.

If, like the local authority in Table 7.4, the building owner requires new buildings for their own retention and use sufficiently frequently as to have set up their own departments to perform the development and building roles, then the combination of developer builder is more accurately described as a 'vertically integrated building owner' (VIBO I). If the roles of developer and designer are combined, then we have either the *speculative designer* case or another version of a 'vertically integrated building owner' (VIBO II). Another version of a partially integrated production system combines the roles of builder and designer and may be called the 'package deal'. However, just to confuse things, this is now perhaps more commonly referred to as the 'design-and-build' variant of the contracting system, especially where the developer acts as a separate client. This can clearly be seen in Figure 7.1, because the

Table 7.4 Actors and roles in a typical new public housing project, Inner London, 1950s/1960s

Roles	Actors			
	(1) Local authority	(2) Sub-contractors	(3) Former landowner	(4) Households
	Supplier of social housing. Internal architects, QS and DLO (building works department)	**Single-trade contractors**	**Property holder**	
Developer	Buys land ahead. Obtains full planning permit for a scheme design. Develops site		Sells freehold of land to (1). Compulsory purchase used for slum clearance	
Designer	Architects' Department designs to meet minimum housing standards and policy objectives			
Builder	DLO project manages the specialist trade contractors. Employs building workers. Buys some materials	Work to contract for (1), on supply-and-fix contract. Employ building workers. Buy some materials		
Owner	Housing Department maintains and operates building on completion			
User				Allocated dwelling on basis of need; rent building on completion

Table 7.5 Actors and roles in typical new office project, City of London, 1980s

Roles	Actors					
	1) Property company	2) Main contractor	3) Sub-contractors	4) Former landowner	5) Professional service firms	6) Financial or commercial firms
	Property holder, e.g. MEPC	A management contractor	Single-trade contractors	Property holder	Design firms	
Developer	Buys land ahead from (4); obtains full planning permit for a scheme design. Develops site			Sells freehold or leasehold of land to (1)		
Designer	Sets design brief; approves design which meets investment criteria				Designs and acts as agents for (1)	
Builder		Construction management or Management fee contractor. Selects and project manages the specialist trade contractors	Work to contract for (2), on supply-and fix contract. Employ building workers.; buy materials			

Table 7.5 Continued

Roles	Actors					
	(1) Property company	(2) Main contractor	(3) Sub-contractors	(4) Former landowner	(5) Professional service firms	(6) Financial or commercial firms
	Property holder, e.g. MEPC	A management contractor	Single-trade contractors	Property holder	Design firms	
Owner	Has ownership rights over the property assets					
User						Rents building on completion. Maintains and operates building (repairing lease)

combination of only designer and builder excludes the client's role as developer.

Systems and their main types of actor

The economic structure of relationships between actors in the building process is profoundly different when the builder is also the developer. For this reason, and also because they are the most widespread in practice, we identify just two main structures of the process, out of the larger number of possible variants: the *speculative* and the *contracting* systems.

In the speculative system one firm combines the roles of developer and builder (and perhaps, also, of designer), but not the role of building owner. Its main application in the UK has been in private speculative house building, though we also find it in some commercial and light industrial building. The main actor, obviously, is the speculative builder.

In the contracting system the roles of developer and builder, and possibly also designer, are separated and performed by different actors. The main actors are the clients, the main contractors and the professional service firms, and there is something nearer to a balance of power between them. Each actor derives power from their role in the development process.

In the Latham Report (1994), clients are seen as the dominant actors, first because the others depend upon them for work, and second because they define the terms of projects. They may also, but do not always, use their position to lead and lay down the terms under which the construction process is organised. On the other hand, their position is weakened by their remoteness from knowledge of building production, and by their inability to specify completely either the product they require or the process by which it shall be produced. Architects and design engineers are the most powerful of the professional service firms, their power arising from their near monopoly over the ability to turn a client's rather abstract functional requirements, expressed in words and numbers, into something concrete and spatial, namely a design, which can be built. In some countries, such as Spain, this monopoly power is supported by legislation, because the signature of an architect is required for planning consent. This power makes them equally necessary as intermediaries to clients and to contractors. Contractors' power comes mainly from their ownership and control of the means or resources of production, and also from their technical knowledge of production processes.

It has been observed that, in the contracting system, the balance of power shifts as each project proceeds. At first, the client is very powerful, especially if their project is a large one, with many competing architects and contractors keen to win the client's approval and selection. However, as the project proceeds and contracts for design and construction are placed, power shifts, and the appointed architect and contractor virtually become temporary monopolists in whose hands the clients have placed themselves more or less irrevocably (Hillebrandt, 1984). This is as true whether we measure their power by the number of competitors, the cost to the client of terminating a contract and reopening competitive selection of a replacement supplier, or the possession of information about the project. Repeat or recurrent clients therefore have a special countervailing power, because *opportunism*, or abuse of their temporary monopoly of supply by architects or contractors, can be kept in check by the sanction of exclusion of those suppliers from future projects.

One way of gaining some insight into relative power within the contracting system is to look at which parties are dominant in which aspects of the construction production process. The key rules of the game of the contracting system concern the following points:

1. How are architects and contractors to be selected by clients? In other words, how is competition to be regulated?
2. How is payment due to each party to be determined?
3. What are the obligations or duties of each party to the contract to the others, and what sanctions can be applied if they are not met?
4. How are changes or events unforeseen at the time the contracts were signed to be accommodated into the project?

For building projects, the normal forms of contract, defining the rules of legal relationship between actors on a project, have for a long time been those devised and published by the Joint Contracts Tribunal (JCT). JCT standard contracts cover the relationship between developer (client) and builder (contractor). The professional bodies, such as the RIBA, produce standard contracts for the appointment of the designer, but these are less widely used in practice (especially by commercial clients) than are the JCT Forms.

In the period up until the 1970s, bodies representing the construction professions and the contractors had larger and more active representation in the JCT than did clients. Public sector clients made little attempt to dominate or to rewrite the rules, though they were more active at the margins of the system in introducing some organisational innovations (see below). Even the main apparent exception, the Banwell Committee Report (1964) on 'the placing and management of contracts in building and civil engineering', seems to have been greatly influenced by representatives of the larger contractors. The standard form of contract of those years, JCT '63, is now widely perceived to have been biased against the interests of clients (Ive, 1995).

Since the 1970s powerful private clients (both owner-occupiers and speculative developers) have been somewhat more active in setting the rules. The British Property Federation developed its own form of contract unilaterally, and many clients now have their own standard variants or departures from the basic form of JCT contract.

Speculative builders and contractors

Speculative builders have distinct economic features compared to contract builders. To begin with, speculative builders initiate work from within the construction industry whereas contractors respond to demand in the form of orders from outside. As a result, speculative builders take on a commercial risk over and above the risks taken by contractors, namely the risk of not finding a suitable buyer for the completed project. On the other hand, it means that speculative builders are in a better position to plan their immediate future activity, and can be more pro-active rather than re-active in shaping their workload.

Speculative builders and contractors are often both subsidiaries of the same large diversified construction groups. The parent companies can use their position and size to provide funding and finance for speculative building projects, and may switch capital between their contracting and speculative subsidiaries. This pattern of ownership can be found in, for example, John Laing plc, which owns contractors, and a speculative builder, Laing Homes. Despite this frequent common ownership, the internal economic characteristics of contractors and speculative builders are quite different, indeed almost opposites, and for this reason are they usually run as quite separate business entities.

The contracting system also gives rise to independently owned professional services firms, including architectural, surveying and engineering professional practices. These firms can be characterised by complete confinement to one role, and by a low requirement for capitalisation relative to the scope and value of the projects on which they work. As yet, in the UK, these professional service firms are not usually subsidiaries of large construction conglomerates, though architecture departments, employing qualified architects, do exist in a few of the larger contractors, such as Kyle Stewart (HBG).

Construction as a manufacturing process

Our earlier definition of actors in the development process tends to downplay the importance of manufacturers of building materials and components. Because the process is seen as starting with the developer, then moving through design and construction, the role of the manufacturer seems to be a long way downstream and remote from the project. However, with reference to Chapter 1, p.10 if we define the process with which we are concerned as the process of production of the built environment, then off site manufacturing, in one form or another, will today normally account for well over half of all that process, measured by value added.

Building can be seen as merely the last, on-site stage of a production process which links a series of industries. This process begins with the taking of natural resources, and involves their progressive transformation by manufacturers into forms, which can then be assembled into buildings. These manufacturers too must be recognised as part of the process. There are, of course, as in other sectors, many firms in manufacturing who sell some part of their output of goods or services to construction firms, but for whom this link is not strong enough for them to be included as an integral part of the construction process. To some extent the cut off point is arbitrary. A manufacturer can be said to have an active role in the construction process either if they target, market and design some of their range of standard products specifically for use in the construction process, or if they respond to commissions to produce products specifically for a certain project.

Other roles in the process

Two important features were deliberately omitted from Table 7.2. They are the roles of merchant and financier, and the various types of actor who perform these roles. These are now examined in turn.

Intermediaries – merchants and agents

It is sometimes helpful to discuss transactions in terms of principals and agents. Principals act on their own behalf, while agents act on account of others. The transactors in Table 7.2 are the principals in the development process, whereas the role of the estate agent, say, is that of an agent acting for a fee on a seller's behalf. The distinction between agent and principal is about who bears the risk of standing to lose capital if things go wrong or of making exceptional profit if things go well. Many consultants exist in the construction industry, who sell professional advice to a principal but are themselves not otherwise engaged as transactors in the process.

The distinction between an intermediary and an actor is that intermediaries merely pass something on unchanged (like merchants) whereas actors produce or use and thereby transform inputs into different outputs. Unlike agents, some intermediaries, like builders' merchants, act on their own account. That is, they buy, hold and sell, and for a while are owners of the commodities they buy. The role of intermediary is made possible only because of the space between the separate actors. For instance, the role of a builder's merchant would be excluded from the building process altogether, if a builder chose to purchase all supplies of materials directly from their producers.

Ball (1988) suggests that it is best to think of the modern building *main* contractor as a kind of merchant rather than as a building producer. Now a pure merchant role would be a simple intermediary buying parts of buildings from their actual producers, such as the specialist, trade or *work package* contractors, adding a merchant's mark-up and re-selling them to building owners or clients. The implied comparison is with agricultural merchants buying, say, grain, from many farmers before selling it on in bulk to food manufacturers. The value of merchants' activities to their customers is that they simplify the process of buying. Merchants provide *one stop shopping* with a single point of responsibility for delivery. Ball was led to this position by observation of the historical process by which main contractors have increasingly detached themselves from direct control over production and from direct employment of productive construction labour. One can see his point.

However, main or management contractors are still, in terms of their economic role, producers rather than just merchants, insofar as they retain just sufficient control over how production is carried out by each specialist contractor and, above all, because they still retain the role of co-ordination of these separate production activities. Nevertheless, one could hardly object to Ball's actual term for them, 'merchant-producers'.

During the 1980s there occurred a shift in the way clients could procure buildings. New procurements systems moved away from traditional methods and introduced, in addition to design-and-build, new management oriented procurement systems (Masterman, 1992). Such management-based systems have variant forms, with the common feature that all production is divided into 'work packages', each undertaken by a specialist firm.

In the construction management (CM) variant, the CM firm provides only management services concerned with planning, co-ordination and the payment of the work-package contractors. Although communication is always through them, the CM firm is not the customer of the work-package contractors, whose contracts are directly with the client. In CM procurement, the CM firm is often introduced at a very early stage, and can therefore affect design. However, the CM firm, because it is not a direct party to the work package contracts, reduces its risk, and project risk lies mainly with the client.

In the management contracting (MC) variant the MC firms offer similar management services to CM firms, but typically of less scope because of beginning at a later stage in the project. Again, payment is in the form of a fee and risk again lies with the client. However, because all the project cash flows pass through the hands of the MC firm, they may share with traditional main contractors the characteristic that their business is cash generating and has a negative working capital requirement.

Construction management and management contracting are particular procurement methods, whereas project management is a form of consultancy, not linked to any particular procurement method. Project managers, acting as agents for their principals, will advise them on appropriate procurement methods. Nevertheless, it is clear that project managers, construction managers, and management fee contractors have increased the number and types of agents already present in the construction process, such as architects, structural engineers, and quantity surveyors.

Financiers of the development process

Almost all actors in the development process are either unable or unwilling to finance all of their activities from their own financial resources. Therefore they borrow from banks and other financial institutions. This aspect of the development process has two sides. On the one side, a significant proportion of all bank lending is made to finance development, including construction, and, on the other, interest payments on loans absorb a significant part of total development revenues. Although the role of lender does not by itself imply any active part in the development process, the presence of lenders is usual at every stage. Banks limit or constrain the behaviour of their borrowers, particularly through their rules on collateral. There are two main forms of financing, corporate finance and project finance, distinguished by the nature of the lenders collateral or security.

Corporate finance

All participants in the production process require finance. The funding is supplied by the financiers to the actors as institutions or firms and not for their particular roles. Once a loan has been made, it is up to the actors to decide exactly how the money should be allocated and spent. The lender, as such, does not stipulate how the money will be spent. Such stipulations would involve the financier in the management of the process. Where this indeed occurs, the financial institution becomes a financier developer. However financier developers in the UK are unusual. Instead lenders concentrate upon the security or collateral for recovering their loans should a borrower default on repayments. This security consists of *all* the revenues and assets of the borrower corporation.

Project finance

Funding is supplied by financiers earmarked for a particular project. The developer sets up a company whose sole asset is ownership of the project (called a Project Company or Special Purpose Company). This SPC then borrows from the financiers, and incurs financial liabilities. Loans to the SPC are thus secured only upon the revenues and capital value of the project itself, for these are the only revenues and assets of the borrower, the SPC.

Early uses of project finance in the UK included some major North Sea oil and gas extraction projects, and (a little later) speculative office developments such as Broadgate. The Channel Tunnel built by Eurotunnel, which was in effect an SPC, was unusual in that from the first its new equity was sold through the stock market to the generality of shareholders. It is more usual for the equity capital of an SPC all to be held, initially at least, by a small consortium of firms who have shared the role of developer of the project.

Project finance limits the liability of the borrower-developer to the value of its equity stake in the SPC. Lenders of project finance will become deeply involved in appraisal of the project's projected costs, revenues and risks. The costs of undertaking this appraisal are reimbursed by the borrower.

Unless a project has exceptionally predictable costs and revenues, project lending is usually riskier, from the perspective of the lender, than lending to a well-established corporation, and this risk is reflected in the interest rates charged. Recent years have seen massive expansion in the use of project finance. It is central, for example, to the Private Finance Initiative.

Projects and firms

Projects are temporary and finite whereas the firms that engage in completing projects are, or at least plan to be, permanent and long-lived. This section explores the ways projects and construction firms intersect one another.

The construction industry often produces extremely large units of output, namely large buildings and works. Indeed, some construction schemes are amongst the largest projects undertaken by humanity. Projects can be large in terms of physical scale as well as costly in terms of money and resources. Consequently production, though finite in time, can take several years to complete. Moreover, as each project passes through its stages towards completion, the content of the work to be done by each firm involved changes. In many cases, specialist firms may be involved for only short periods during the total production phase of a project.

Occasionally construction firms are created specifically to perform on one project, and then liquidated. Usually these are consortia or joint ventures between several more permanent firms. Trans Manche Link, for instance, was an important consortium of major contractors formed solely to promote and then to construct the Channel Tunnel.

In many ways the project orientated features of the firm are similar in both the construction industry and in ship building. Because of the size of construction projects and the relatively few building projects undertaken by a firm at any one time, each contract will form a relatively large proportion of a construction firm's turnover. Consequently, each contract forms a significant opportunity for the firm to develop and grow. However, each contract is also a threat to the firm's survival, if the project in the agreement were to run over budget.

When construction firms are very small, it is common for them to obtain all or most of their turnover in any short period from just one project. As firms grow just a little larger, they tend to do so by running work on a few projects in parallel, where each involves roughly the same turnover for the firm as the single projects of the smallest firms. Sometimes, in the next phase of expansion the firm enlarges its scale of operations by increasing the volume of work on each project it is willing to undertake. It raises the amount of revenue before expenses per project. In other words it increases the size of its turnover per project. Instead of increasing the number of small jobs and spreading management more thinly, firms expand by running the same number of jobs but larger, with approximately the same number of senior managers. Although it is larger, the firm still depends to a large extent on the success of each project. Thus growth proceeds, first incrementally, by adding more contracts of an accustomed size in parallel, then by quantum leaps to a higher scale of project.

It is noticeable that the biggest contractors, for example, do not seem to have noticeably more projects under way at any time than firms a fraction of their size. They are bigger, rather, because their average size of revenue per project is bigger – and that is either because they work on projects with a bigger total size or because they take a larger slice of the projects.

Data on main contractors' value of contracts per project can be found in *Housing and Construction Statistics*. Unfortunately, it is not possible to link it directly to contractors' size, i.e. total orders obtained per firm, but it is possible to see that the size distribution of firms closely follows the size distribution of new build orders. However, the conclusions which can be drawn from Figure 7.2 can only be tentative. While the percentage of the number of orders in each project range reflects the percentage of firms in each size category, the percentage distribution of the value of work is almost the reverse, with the largest firms obtaining over 40 per cent of the work. Figure 7.2 as an example shows orders for industrial new work. It does not take repair and maintenance contracts into account, and maintenance contracts form almost half the workload of the industry and are predominantly small in value and undertaken by small firms. If repair and maintenance projects were to be added to Figure 7.2, we might find a more or less constant ratio, in each size range, between the proportions of the number of firms and the number of contracts.

Subcontractors, or direct specialist contractors, on the other hand, tend to work on many more projects in the same time period than do main contractors of comparable size. This reduces the exposure to the risk of bankruptcy of small contractors, because they are less dependent on each project for their survival. For subcontractors, the average ratio of turnover per project to the firm's total turnover is much lower on average than main contractors.

Similarly, amongst professional practices, firms of quantity surveyors tend to have more projects in hand at any one time than architectural

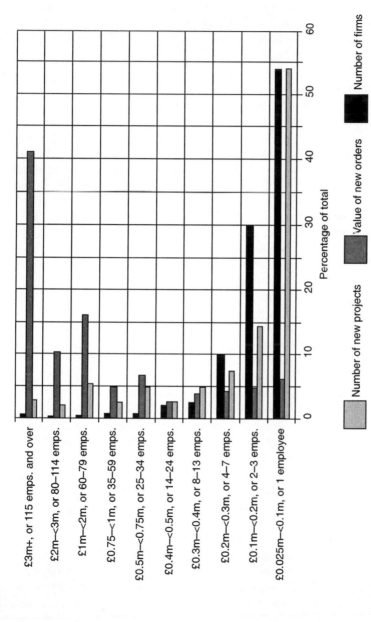

Figure 7.2 Percentage distribution of firms of main contractors, by number of employees, and number and value of industrial new work contracts (payments), 1997

Source: Adapted from *Housing and Construction Statistics, 1987–97*, Tables 1.4d and 3.1a (London: HMSO).

firms with comparable annual total turnovers. Quantity surveying practices have lower revenues per project than architects, reflecting their lower percentage fees, and consequently they require more jobs in order to achieve a similar total overall revenue.

Average and median project value varies significantly between different project types. In 1988 just prior to the last major recession in the construction sector, for instance, 47 per cent by value of private commercial construction contracts had an individual value of £5 million or over, whereas less than 2 per cent by value of housing projects, public or private, were of this size. This compares to 20 per cent by value of private industrial projects, and 33 per cent by value of public non-housing projects. In terms of numbers of projects, of course, the vast majority in all types are very small, but, by share in total market value, individual projects of under £0.2 million accounted for only 15 per cent of private commercial, 18 per cent of public non-housing, 19 per cent of private industrial and 37 per cent of private housing. In a recession year the share of large projects is typically somewhat less, at least in the private sector. All these figures refer to new build only. They exclude projects classed as repair and maintenance, where project value is presumably much smaller.

Figure 7.3 shows the distribution of contractors' new orders in 1997 by project duration. Over 80 per cent of orders take up to 6 months to complete whereas only a few involve projects lasting one year or more. This of course is in contrast to the value of these orders, since projects with longer duration tend to be larger and more valuable than the shorter jobs. This contrast is illustrated in Figure 7.4, which shows the distribution of orders by duration and value.

During the 1980s several major projects in the UK received much publicity due to their speed of construction. The management techniques involved were called 'fast track'. This is a management system which enables the construction phase to overlap the architects' drawings stage, so that the overall development period could be shortened. This was achieved by dividing projects into overlapping 'work-packages', and by having regular meetings on site between clients, architects, contractors, subcontractors and engineers. One notable example of the use of fast tracking was the Broadgate Centre in the City of London.

If there has been a tendency in recent years towards faster construction – that is, shorter contract durations for projects of equivalent real size – it is difficult to measure the effectiveness of the techniques involved from published data. Nevertheless, faster building has been a prominent recent concern of major clients.

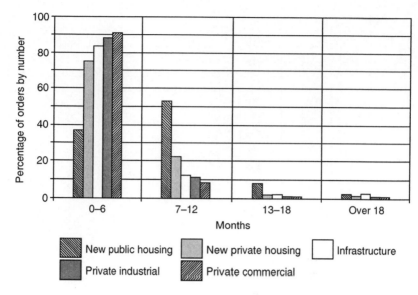

Figure 7.3 Contractors' new orders, by duration and number 1997
Source: *Housing and Construction Statistics*, 1987–97, Table 1.5.

If project and contract sizes stay unchanged whilst durations are reduced, then several outcomes for construction firms are possible. For example, it could make it harder for construction firms to plan ahead by predicting future resource requirements from their current order book (Sugden and Wells, 1977). It could mean that a firm will be working on fewer projects at any one time than before, but with a higher monthly value of turnover from each project. If the number of projects also remains the same, it could mean that following a restructuring period when firms merge or close down, the remaining individual firms would be able to work on the same number of projects at any one time as before the 'speed-up', with a higher annual turnover. Firms would then be able to take advantage both of economies of scale and new computing and communications technologies.

The risks and uncertainty connected with building projects can be enormous. Contracts are signed before construction begins and thus before actual costs are known, in contrast to much of manufacturing industry, where goods are offered for sale only after production has taken place. A second major source of uncertainty for contractors is that it is particularly difficult in construction to predict the costs of one project from past experience of similar projects, because it is hard to

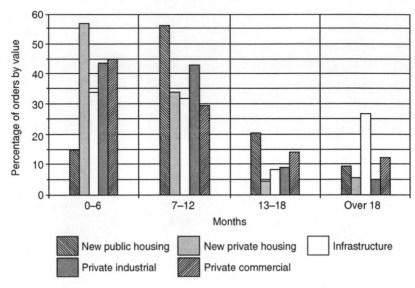

Figure 7.4 Contractors' new orders, by duration and value, 1997
Source: *Housing and Construction Statistics*, 1987–97, Table 1.5.

know in advance just how similar any two projects are going to turn out to be.

Portfolio risk management costs

Portfolio risk theory has been developed mainly in the theory of finance. There, it has enabled identification of *optimal* portfolios of financial assets, including equities in different companies plus bonds. It has shown that holders of such portfolios will always be at a considerable advantage over holders of single assets, or less diverse portfolios. *Inter alia*, this provides an explanation in efficiency terms of why modern holding of financial assets is dominated by financial intermediaries, such as insurance companies, pension funds and unit trusts, rather than by direct individual share holding. The value of an optimally composed portfolio of, for example, shares will fluctuate less than will the value of any other kind of holding. Optimality requires, simplifying somewhat, that the portfolio not be imbalanced by too large a holding of any one asset, and that it be composed as far as possible of assets whose values tend to vary independently of one another.

For example, they should be shares in companies trading in different industries and markets.

Now, the same ideas have a clear application to the portfolio of projects or contracts held at any one time by a construction firm. The value of the outstanding cashflows from each contract in a firm's portfolio of jobs in hand will be prone to vary, in part for reasons specific to that project and unlikely to affect the value of the other contracts in the portfolio. Because of the independence of this variation, the larger the value of the portfolio relative to the value of each single contract in it, the less will be the mean expected variance in the combined value of the whole portfolio. Starting with a small portfolio, if we double its value by doubling its spread in terms of numbers of contracts, we will observe something much less than a doubling of the absolute size fluctuation in its combined value, as each project is worked through to final settlement of accounts.

Now, portfolio theory normally develops its argument in terms of risks of known probability, rather than in terms of uncertainty or uninsurable risk. Whatever may be the case for the stock exchange, where *futures markets* do permit some *hedging* of risk, when we consider portfolios of contracts held by a construction firm, it seems fairly self-evident that most of the risk involved is not capable of being laid off or insured, though of course certain specific categories of contract value risk can be insured, such as the Export Credit Guarantee Scheme, certain kinds of contract bonding, etc. At any event, even if insurance might be hypothetically obtainable at some price for the remaining risks, the judgement of firms seems clearly to be that the premiums payable would not be good value for money.

Thus construction firms themselves carry the risk that the remaining value of any contract, C, in their portfolio might become negative with future cash outflows exceeding cash inflows. They manage this risk not chiefly by taking out insurance, but rather by holding contingency reserves of relatively liquid financial assets in their corporate treasuries.

Suppose a firm has a portfolio of 20 contracts, the expected *ex ante* outstanding net cash flows of which range from – £6 million to +£10 million, and the expected sum of the net cash flows is £40 million (note that these are net cash flows, i.e. expected revenues minus expected outgoings; the gross expected cash inflows, the value of the order book in terms of turnover, would be very much larger).

If each contract were to turn out as expected, the firm would have no need to inject cash from its treasury into the contracts division – each month the positive cash flows from the majority of projects would more than cover the negative cash flows from the rest.

However, whilst 19 out of 20 contracts may indeed have outcomes that are within a few million pounds of what was expected, the problem for the firm is that experience shows that there is a significant chance that 1 out of the 20 will run into major unexpected problems. – let us say, for purposes of exposition, a negative cash flow of £40 million spread over several months.

Now the ordinary monthly inflows from the cash generating projects will be insufficient to cover the cash outflows in several months, even though overall the firm is still just about breaking even. The disaster loss just about offsets the expected sum of net inflows from the whole portfolio. It is to cope with such disasters that the firm needs its liquid financial assets contingency reserve.

Experience will indicate to the firm the minimum prudent size of such reserve that it must hold, in order to feel *virtually* certain that it will not be caught unable to pay its creditors in the event of a worst case scenario on one or more of its projects.

The contingency reserve might take the form of open but undrawn upon lines of credit with its bankers, or might take the form of readily encashable holdings of financial assets such as bank deposits, government bills or bonds or equities. The cost of having to keep such a reserve may be the opportunity cost arising from the fact that the rate of interest on such holdings will be less than the rate of return the firm could make if it could release that money capital for real investment in its operations, or, if the reserve financial assets have been financed by long term borrowing it is the actual excess of the cost of servicing this debt over the interest obtained from short term deposits.

Now, the size of the worst case scenario negative cash flow from the portfolio will depend upon the following variables:

- the minimum quality of the individual contracts. Contracts with some clients are of lower quality considered as financial assets than others, because attributes of those clients mean they are more likely to default on payments. Such clients might include highly geared speculative developers;
- the maximum technical uncertainty of the individual projects underlying the contracts;
- the extent of liabilities for damages implicit in the value and form of contract.

Generally, forms of contract are such that the absolute magnitude of the worst case negative net cash flow that can be imagined is in fact directly proportional to the gross value of the contract. Thus, other things being equal, the larger the absolute value of individual contracts relative to the size of the firm, the larger the negative cash flow of the worst case scenario relative to the ordinary net cash flows of the firm, and therefore the larger the contingency reserve required as a proportion of the firm's capital employed.

Concluding remarks

There are a great number of separate participants involved in the development and construction of any project. While all participants in a project have their own motives and expectations, it is the developer who has more power than the others to determine what actually gets built. In the main the participants or actors belong to separate firms or organisations. It is perfectly possible for them to be vertically integrated, such that different roles may be combined within the same organisation.

In many ways projects can be seen as power struggles between participants. These power struggles are conditioned by the contractual arrangements made for the construction and delivery of the project. The client has the dominant role in this process and can decide how the rest of the project participants are selected, as well as their terms and conditions of employment.

In spite of the high degree of uncertainty surrounding any project, contractors find themselves competing for work on very low profit margins. They are thus exposed to a high level of risk, albeit not 'demand' risk. In order to reduce the risk of bankruptcy, contractors work on several projects simultaneously. This method of managing their workload reduces the risk that any one project may cause the company to fail. A portfolio of projects is used by contractors in much the same way as a portfolio of financial assets is used to off set losses in one investment with profits from the others.

8
The Contracting System

Introduction

To some extent the way that any industry organises itself is a collective response by the firms of that industry to their economic, social and political environment. Firms also need to take appropriate measures to deal with the specific types of inputs, the nature of the technology and methods used, and the characteristics of their output. For this reason every industry is peculiar to itself. As industrial economists we seek to understand the nature of the conditions which determine the behaviour of firms. One of the main characteristics of the construction industry is the division of the production process between separate professional design practices and construction contractors.

This chapter deals with the implications of that division in terms of the ability of the industry to introduce improvements and increase productivity and efficiency. We see how the current contracting system operates to the detriment of clients as well as of the long term interests of the industry itself. And yet, given the current terms and conditions in construction markets, for any one firm to attempt to circumvent the contracting system on its own would force it to tender at uncompetitive rates and so risk bringing about its own collapse. Those contractors which operate most effectively use the tendering system to win orders, while using the subcontracting system in turn in their own interests to cut costs. This chapter focuses on the relationship between the contracting system and project innovation.

Fragmentation and innovation in the development process

Fragmentation between design and construction reduces efficiency and raises the costs of production. A result of fragmentation has been the low level of investment in research and development within firms in the construction industry (Ive, 1996).

We have seen how the organisation of construction firms emerges from the financial and legal constraints operating on companies. The organisation of the industry itself is also a consequence of historical, legal and financial developments. Partly because of the uncertain work flows resulting from the tendering system and partly because of the low profit margins, in the construction industry the strategy for survival of firms has led them to cut overhead and fixed costs to a minimum. This has meant, for example, the growth of a plant hire market as firms prefer to hire than own the plant and equipment. Subcontracting also means that specialised skills and equipment are not a charge to the main contractor firm when they are not required on a site.

In the professions the sector is represented by a large variety of institutions including the Royal Institute of British Architects, the Royal Institute of Chartered Surveyors and the Institution of Civil Engineers. Each of these professional bodies represents the interests of its members. This preserves the separate identities of the professions and means that the professional practices often cater for only one profession. Design teams have to be set up anew each time a major building project is planned. At the same time each project requires contractual arrangements to meet the unique needs of the organisation created to design and construct the scheme. No universal contract applies.

Ball (1988) accepts that technical change has made it possible to create buildings in ways that were previously not practical. For Ball, the issue is why does technical change and development in construction take the form it does? The best theoretical framework developed to date to study innovation in the UK's vertically fragmented construction process remains that of Marion Bowley (1966). Below we offer our interpretation of her ideas, and of major subsequent additions to them.

Construction in the UK is dominated by *the contracting system*. The twin essences of this system are the separation of the roles of project initiator, designer and producer and the one-off nature of construction projects. Each of these is inimical to innovation. The separation of the roles divides responsibilities between the client, the architect and the contractor, making the introduction of new techniques, technology and materials difficult to implement. Construction demand is seen as a series of *one-off* projects, each with its unique design, tendered competitively on price, and with its own temporary project coalition. Ball (1988) and Winch (1989) develop Bowley's point. Ball argues that because each construction project is a one-off and it is not possible to predict the next project, contractors need to be highly adaptable or flexible in their ability to respond quickly to the specific requirements of each contract, rather than focusing upon efficient delivery of the kind of project currently in hand.

Innovation is to be distinguished from invention, as requiring development beyond the *prototype* or *experimental project* along a learning curve and towards routinisation and eventual diffusion through imitation. Innovation, like murder in a properly made 'whodunit', requires the coincidence of means, motive and opportunity in the same party or set of allies.

Bowley assumes the desire for temporary super-normal profits is the main motive for innovation, whereas the desire to try out a new idea, or the search for fame and attention may be the motives for invention. *Normal* profits are the rule, but an innovator can obtain super-normal profits for so long as their lead over competitors lasts, which is until diffusion becomes widespread, and the innovation becomes the new norm.

Innovations are of two kinds, product-enhancing or cost-reducing. The former involves product innovation, the development of what customers perceive to be a *superior* product. The latter involves process innovation, new ways of making what is perceived to be the *same* product. Super-normal profits can arise from the ability of the firm to use an innovation in order to raise its selling price or expand its volume of sales. These possibilities result from *product enhancing innovation* which enable firms to raise prices or to capture market share and take advantage of economies of scale. The options derived from *cost reducing innovation* are to cut prices and thus expand the firm's volume or to widen its margins if cost reductions are not passed on to customers.

The motive to innovate therefore depends upon there being a substantial time lag between the introduction of an innovation and it becoming the *new norm*, and upon the size of the extra profits to be made in each year of that lag period. These extra profits depend upon the percentage difference an innovation makes to cost or price and the total value of projects on which it can be applied by the innovator. These concepts are known as the *magnitude* and the *scope of application* of an innovation.

The contracting system, according to Bowley, means that architects and, especially, contractors only have the *opportunity* to attempt innovation *after* they have been appointed by a client – they can only be re-active, not pro-active. Clients, on the other hand, have this opportunity to innovate from the moment they conceive a need for a project. It means also that architects and contractors partially lack the *motive* to innovate. In both cases, this is because they cannot know in advance that they will be able to use their innovation on future projects, because they cannot control what these will be. It is also because architects have no direct profit to gain from cost reducing innovation within the initial project. Contractors, on the other hand, have ample direct motive to make cost reducing innovations, though no real interest in product enhancing ones, but lack the opportunity to do so, because most process innovation actually requires at least minor project redesign, which is outside of contractors' control. Architecture firms, being under-capitalised and unable to carry their own uninsurable risk, also lack the financial *means* to invest in innovations.

Whether or not this conceptual framework derived from Bowley is a *sufficient* base for thinking about innovation in the modern construction industry depends in our view largely upon two aspects of the development process. It depends firstly, on the importance we place on horizontal as opposed to vertical fragmentation in the process, and second, on the importance we place on following the process *back* below the contractor to subcontractors and manufacturers. Let us consider these points in turn.

Bowley's explanation of the relatively slow rate of innovation by firms employed in the construction process is meant to hold even if there is complete horizontal integration of each stage on each project. Complete horizontal integration would mean only one client, only one designer and only one contractor. However, it is really impossible to grasp the forces at work behind recent changes in the contracting system without reference to horizontal fragmentation in both the design and building processes, and the consequent rise of *co-ordinators*. By this term we mean to encompass all those '*project management*', '*construction management*', '*management contractor*' and '*design management*' actors whose role is not directly to design or construct, but rather to *manage* and co-ordinate the activities of other actors amongst whom these activities are divided. *Management* here does not have its normal meaning, of direct employment of and hierarchical authority over those being managed. Instead, it means something nearer to procuring or selecting, monitoring and co-ordinating.

Why has horizontal fragmentation increased? Partly because some projects have become much more complex and large, and partly because of the general *economies of specialisation* first observed by Adam Smith. Specialist firms have the ability to carry the division of labour to a further degree. Most main contractors have followed a policy of minimising their specialisation in terms of client or product markets, in the sense of confinement to or reliance upon one type of client or one type of product. They have instead marketed themselves as *generalists*, capable of constructing anything for anyone (except perhaps for a specialisation on projects above a minimum size).

In contrast to main contractors, most subcontractors have followed the opposite course, and have specialised on one very specific type of construction activity or *trade*. Demand for even highly specialist trades, like suspended ceiling installation or lift installation, may be more stable than demand for a particular type of building or works, such as social housing or hospitals. Some main contractors do nonetheless specialise by product type (tunnelling contractors and motorway builders, for example) but this is because these markets require very expensive, specialised technology and equipment for firms to be at all competitive in them.

Horizontal fragmentation of production between many specialists reduces clients' effective direct power over the process, and therefore their ability to force through innovation, in three ways. First, it results in extra layers of organisations of designers and builders intervening between clients and the actual producers. These extra layers extend lines of communication and reduce contact, knowledge and power to negotiate. Second, it may make each producer engaged on a client's project less *dependent* on that client. The fragmentation of work reduces the ratio of any one specialist producer's contract value to that producer's turnover. Moreover, the specialisation of the service provided reduces the likelihood of repeat work for the same client. Third, it increases the number of firms with which an innovating client must establish altered relationships, and hence the time and effort the client must expend in innovation.

It also means that neither the design or construction co-ordinators nor the specialist designers and constructors are in a position to introduce wide ranging innovation on their own initiative. The co-ordinators may wish to achieve process innovations, but are perhaps too remote from actual involvement in the practical aspects of production methods to do so effectively, other than by exhortation. Whilst a specialist is able to innovate only if it has no knock on consequences for technical aspects of the project outside of the boundary of their own *work package*.

We turn now to follow the process back to manufacturers. Where do we locate manufacturers within a *contracting system* type of analysis? Bowley assumed that clients have no direct contact with or knowledge of manufacturers, and that manufacturers' relationships are entirely with designers, specifiers and installers. There is also assumed to be no vertical integration between manufacturers and builders. Design of components is seen either as a mass production process undertaken by manufacturers, or a top-down process led by project architects and engineers. In the Bowley framework the key distinction is between manufacture of *specials* and *standards*. These ideas pre-date the emergence of flexible manufacturing, assume that standards will always be much cheaper than specials, and that the batch size needed to exhaust economies of scale in production is large. All this needs to be updated in the light of the *revolution* in manufacturing techniques. As noted above, production has been transformed to some extent over the last two decades by an increasing use of robots and programmable machines in factories.

The role of *the client* in relation to innovation in *the contracting system* depends on whether clients adopt either a strategic or a tactical approach to their role in the development process. If clients adopt a strategic approach to their role, they, uniquely, are in a position to rewrite the rules of the game. At least, they can devise new forms or variants of the system, which we call procurement routes. If they choose to adopt a tactical approach to their role, they are then enmeshed within the given constraints of the system in the same way as other parties.

Clients may have the *motive* to achieve product-enhancing innovation, especially if their projects are large or part of a continuing programme, and they may wish to achieve cost-reducing innovation if they can be sure to capture some part of that lower cost as a lower contract price. They certainly have the *opportunity*, in the sense of presence at the initiation of a project. However, they may lack the *means*, in terms of the technological capability to identify viable and feasible innovation possibilities. They should know, who better, what innovations would be desirable, but may not have sufficient knowledge of what is achievable. However, it is at least arguable that many clients only *know* they want something new when they have it presented to them (Ive, 1995).

Our discussion of the clients' role in the process concentrates upon their role in stimulating or initiating change. Clients can do this if some clients exist for whom commissioning buildings and works is a sufficiently crucial part of their business to give them the *incentive* to spend the time, money and effort required to lead both the strategic *redesign* of the construction process as well as the tactical process of project innovation. Thus, whereas the standard textbook treatment of *the client* focuses upon a static treatment of the question of *what procurement route* from among an existing and given repertory the client should choose, we ask the question, 'which client led changes to the rules of the *classical* version of the contracting system would most encourage project innovation?'

Their *incentive* to innovate depends upon the importance or otherwise for clients, for their own success as organisations, of commissioning buildings and works, and this in turn seems to break down into:

1. The value of their expenditure on construction investment and operation relative to their total investment and to their turnover. This is largely an industry determined variable, i.e. higher for the *average* firm in some industries than in others, but will also vary significantly within industries, yielding clients who are *leaders* in this respect within their industry. Industries with high averages for building or construction work include retailing, the utilities and process manufacturing.
2. The absolute value of their construction commissioning over an n-year period. Expenditure on innovation partakes of the nature of an indivisible and lumpy fixed cost, only economical if it can be spread and recouped over a large value of construction programme.
3. The degree of similarity between projects comprising their commissioning programme. The greater the proportion of their programme to which any successful innovation could be applied, the greater the return and incentive to spend on innovation.
4. The use of built facilities to give competitive advantage, over rivals in the same industry or over possible substitutes. One can apply the standard project trinity of cost, time and quality to break down the impact of construction on competitiveness. One could distinguish between clients for whom competitive advantage comes through cheaper built facilities and those for whom it comes from faster construction times or more timely construction completion or from better or more reliable quality built facilties.
5. Degree of dissatisfaction with the construction process as it exists. The Latham Report (1994) discusses client dissatisfaction with construction quality, reliability, and value for money. Major procurers, including Slough Estates, Lynton, Stanhope and McDonald's

Restaurants, are seen as examples of client innovators in the construction process, developing the level of specification, programmes of work, and the use of standard components and modularisation.
6. Degree of originality and complexity, and hence of risk and uncertainty, in their project requirements and briefs.

Alongside incentives to innovate we must also consider availability of the *means*. Here, this largely refers to the degree of in-house expert knowledge and the strength of the commissioning function in the client's overall management structure. There seem to be enormous differences between apparently similarly placed clients in this respect.

One theme must be *client competition versus client co-operation* as routes to innovation. Innovation by any one firm or organisation may generate benefits for that firm or organisation but none to any other party. Such gains are called private returns. In the 1960s, it was recognised that often no one public sector client commissioned enough similar construction to yield an economical *private* return on fixed expenditure of time and effort, as well as direct higher initial costs on early phase product development innovation. The solution then tried was co-operation, and the forming of consortia of the commissioning and design departments of many public authorities.

Currently, the dominant private, corporate clients are less inclined towards co-operation with other clients, who may be potential rivals. Nor is government inclined to cajole or push firms into such collaboration. In UK construction, CIRIA represents about the nearest we now get to co-operative sharing of research and development expenditure and benefits between firms. Its scope is limited and clients play a limited role within it. Competition provides a spur to innovation, but with the caveat that innovation will only be appraised on its *internal* returns, which are the private financial returns of the innovator. The Movement for Innovation (M4I), however, represents an interesting recent attempt to 'share' knowledge of some kinds between innovative clients and their suppliers.

Given the diagnosis of fragmentation and separation of roles, responsibilities and knowledge as the roots of the problem of slow innovation in construction, there are two alternative ways of seeking to overcome it.

One is vertical integration to bring the organisations performing separate roles under common ownership, and therefore under a common set of goals and objectives. Examples would include design-and-build firms, manufacturers owning the subcontractors who install their components, and clients owning their own project management and design organisations.

Another way to encourage innovation in construction would be to improve communication and sharing of responsibility between organisations within an overall hierarchy of project control, although the firms and organisations themselves may remain in separate ownership (and therefore related only contractually on a project). In this way, any party to the process, including the client, may seek to extend their influence beyond the boundaries of their own organisation and its in-house activities. This is the domain of the construction literature on 'partnering' (Barlow *et al.*, 1997).

Client roles in establishing integrated information systems amongst stable networks of suppliers, or amongst temporary members of a project coalition, are potentially crucial. The client is uniquely well placed to lead, and to 'knock heads together' when necessary. Information technology (IT) has the potential to transform the density, extent and timing of interactions between clients, designers, builders, and manufacturers.

The contracting system

Since Bowley wrote about the *classical* form of the contracting system, it has been replaced, to a large extent, by new *procurement routes*. In the traditional or classical system the client began by choosing an architect, normally without arranging a formal competition between potential suppliers. The architect was paid a fixed fee, set by the RIBA as a percentage of the value of the construction contract. The architect then, having discovered the client's requirements and budget, developed a complete design. This was then costed by a quantity surveyor, also appointed by the client. In other words, the estimated likely price of construction was checked against the client's budget before proceeding to the next stage. Then, the detailed design, together with the bill of quantities, were sent out to a set of would-be contractors (called tenderers) to be priced. The opportunity to tender could be *open* to all or *restricted* to a set of contractors chosen by the client's agents. Each potential contractor submitted a bid or tender – an offer to complete the works described in the documents for a lump sum price chosen by the tenderer. The construction contract was then awarded to the lowest priced tenderer.

It is clear that if the contracting system works in the way described above, then the contractor can have no influence upon the design, which is completed before they are involved in the project, and like-wise that the architect can have little involvement in construction, other than to check the quality and accuracy of the contractor's work. The two main advantages of the traditional system to clients are supposed to be, first, that they obtain the impartial professional advice of the architect and, second, that they offload all risk concerning cost and price of construction on to the contractor from the moment the contract is awarded. In principle, therefore, the clients should know the out-turn cost of the project before committing themselves to it.

In practice, whether or not it had ever worked like that, the system certainly did not do so by the 1960s and 1970s. Instead, it had become commonplace to go out to tender with incomplete designs and specifications, partly because clients were in a greater hurry, and partly because of the increased complexity of design and the fragmentation of design between various specialists. As a consequence, and also because contractors did not wish to take the risk of forecasting inflation during the life of the construction project, the tender prices submitted had become far from invariable fixed prices for the whole project. Instead, the unit rates attached by the contractor to the items in the bill of quantities became the basis for a continuous negotiation over additional payments to the contractor for variation or additions to the work done from that specified in the tender documents. Likewise, the tender price had also become merely the *base price* upon which would be added extra percentages to compensate the contractor for general increases in wage rates and materials prices.

Consequently, clients became somewhat disillusioned with the supposed benefits to them of having a 'lump sum' tender price agreed in advance of construction. At the same time, clients were increasingly aware that more effective design solutions could affect the total price of a project more than a system that relied on the competitive reduction of contractors' mark ups on pre-determined designs. Developers were therefore willing to listen to contractors' wishes to be more involved in design decisions. Two main *families* of *solutions* evolved – 'management' contracting, in its broadest sense, and 'design-and-build' contracting (Masterman, 1992).

In management contracting, the client appoints a 'managing' contractor or 'construction manager' and pays them a percentage, formula-based or lump sum fee. This contractor then acts *on behalf of* the client, to select, and agree tender prices with, specialist trade contractors. In some versions, contracts with specialist contractors are made with the client, and in others the contracts are with the *manag-*

ing contractor. The client pays the managing contractor as work proceeds, and the managing contractor then pays the trade or work package contractors. Instead of a *firm price* the managing contractor promises to use their expertise to obtain *the best price* for their client, given the usual trade-offs between price, speed and quality of work. In most versions, detail design can be run in parallel with construction. There is no presumption of complete prior design. Construction contracts for parts of the work can be let as and when design of those parts is completed. Alternatively, parts of the design can be devolved to the specialist contractors. The managing contractor may also be responsible for managing the flow of design information from the design specialists to the construction specialists.

In design and build, the client appoints a *representative* to help them develop their ideas for the project to the point where they can be put out to tender to a competition of design and build contractors. These tenders will differ from one another not only in price but also in the proposed design. For a lump sum, the design and build contractor agrees to provide a building or works to the specifications developed before the tender. In principle, the client could simply produce a written performance specification, and leave all design to the design and build contractor. In practice, normally design is developed quite a long way before appointing the contractor. In effect the design and build contractor becomes responsible for detail design. Moreover, the normal practice is then for the contractor to sub-let the production of design details to separate firms of architects, and indeed to sub-let construction to specialist trade contractors. It has, indeed, been said that instead of 'design-and-build' contracting, with its implication of integration, a more realistic term might be 'manage-and-manage' contracting, to stress how it remains within a system of fragmentation.

Concluding remarks

Now, each of these relatively recent variants on the contracting system to some degree reduces the previously complete separation of roles and responsibilities of design and construction. However, the other aspect of the contracting system mentioned earlier, that each project is a *one off* with its own unique and temporary project coalition of actors, remains essentially unaffected by whatever procurement route is chosen. The key question is, are these new procurement systems merely modifications of the contracting system or do they, and particularly does design and build contracting, constitute a fundamental shift to another system of relationships altogether? Bowley's proposed alternative to the contracting system, after all, was a form of design-and-build or 'package deal' (Bowley, 1966). We cannot be sure how design and build contracting will evolve in the future. However, at present the weight of evidence is that these have been *reforms* rather than *revolutions* in the *system*.

In spite of the effort to find a satisfactory procurement method, from the clients' point of view the construction process has remained stubbornly confrontational, prone to late completions, and poor quality. Many of the issues confronting the construction industry including procurement methods, were addressed in the Latham Report (1994). However, arguments between the participants continue. Main contractors continue to use *pay when paid* clauses with subcontractors, causing great contention between the parties to many construction projects. Litigation is common. The industry remains fragmented with the protagonists continuing to defend their particular sectional interests (Gruneberg, 1996). To begin to gain an understanding of the causes of conflict in the construction process we discuss the *transaction cost approach* to market relationships in the companion text, *The Economics of the Modern Construction Firm*.

Part IV
Construction and the Economy

9
Construction Investment, the Multiplier and the Accelerator

Introduction

This chapter discusses construction output and demand in relation to gross domestic fixed capital formation (GDFCF). This approach views construction output as investment in the productive capacity of the economy and hence as a contribution to economic growth.

The official statistics published in *Housing and Construction Statistics* break down demand into markets according to client, activity, product and region:

- The type of client is divided between purchasers in the public and private sectors.
- The type of activity involves new construction or repair and maintenance.
- The type of product is analysed according to its economic function such as housing, industrial buildings, commercial buildings, public buildings, infrastructure. Breakdowns of type of product are further disaggregated into sub-types. Commercial output is divided into offices, shops, garages, entertainment, and private education. Infrastructure includes roads, harbours, water, sewerage, railways and airports, gas and electricity. Public buildings comprise schools, universities, hospitals, health centres, and government offices. While data on these is available on a quarterly basis, industrial demand data by individual industry is only available annually, from the Census of Production for each industry and from the *Blue Book*.

- Finally, national demand, and hence construction output, is divided into its regional location. It is implicit that variations in output in turn determine variations in employment.

Unfortunately it is no easy matter to measure changes in construction workload which indicate changes in demand on resources, even from data collected by government. Nor can official statistics be relied upon to give an accurate picture of the value of contracts in hand. There are several difficulties in interpreting the statistics. One such difficulty, as Lewis (1965) has pointed out, is because of the long duration of building and construction projects, often lasting several years between start and completion, especially on large schemes. The value of construction contracts let in a given year may have little bearing on the amount of work carried out in that year. Similarly, the amount of work done in a given year may also under or over state the quantity of construction work in the pipeline (BERU, 1974; Sugden and Wells, 1977).

There are also problems in converting data measuring orders or demand by *value* into *real* data, which measures volume. It is only from the real data that demand for resources can be calculated. The problems here mainly concern the reliability of the price indices used and the possible variability of the technical coefficients linking input and output quantities.

However, from the point of view of construction firms themselves it is value rather than volume that provides the prime basis of calculation. They are in business to make profits, and profits are a monetary concept concerned with value.

The amount of construction activity varies from year to year and one of the main difficulties facing firms in the construction industry concerns the timing of changes in anticipated future levels of demand. Each firm hopes to win a certain share of the expected volume of work available. If a firm's market share is stable, then its level of sales will depend mainly upon fluctuations in the level of demand in its markets. In some industries, certain firms have significant competitive advantages over their rivals. These advantages enable them to gain market share, and detach their own level of output from overall market demand.

However, in the construction industry, this kind of major competitive advantage is quite rare, and most firms rely on benefiting from market growth to achieve business expansion. Partly as a result, firms that are ambitious to grow come to stress their ability to identify and enter currently relatively fast growing sub-markets, within the overall or aggregate construction market. Thus a kind of 'nimbleness' in shifting between markets comes to take the place of sustainable competitive advantage in any one market.

Moreover, if the firm is to plan its use of resources, it needs to be able to predict the level of demand in time to make the necessary investments or disinvestments. Accurate predictability of demand is in itself important, quite aside from the importance of the actual level (high or low) of out-turn demand. If firms cannot or do not attempt to predict demand, they will instead turn to other strategies involving less investment and less planning ahead – relying on quick response rather than on anticipation.

To aid predicting demand for construction it is necessary to examine the mechanics of the relationship between construction and the rest of the economy. Economists have found two concepts to be of particular value in understanding how construction activity is affected by and in turn affects demand in the rest of the economy. These concepts are the multiplier effect and the accelerator principle, both of which are discussed later in this chapter.

Aggregate construction demand

In the long run the UK economy experiences a pattern of growth, with some years exhibiting higher rates of growth than others. A mild recession occurs when an economy grows at a slower rate than its average rate of growth. A strong recession occurs when the economy shrinks, when GDP is lower than in the previous time period. During recessions unemployment rises, output growth slows down and firms begin to de-stock. They reduce their inventories of circulating capital goods. As business confidence in the market begins to fail, firms also pull back on investment decisions.

Construction demand comprises a mix of consumer demand and investment demand. Consumer demand and investment demand both respond to changes in the economy which they in turn influence, but in very different ways. It has often been observed that investment demand of all sorts changes proportionately more from year to year than does consumption demand. Thus industries producing consumption goods and services face less instability of demand than investment goods industries, which produce fixed capital goods. Construction, together with certain engineering industries, makes up the investment goods branch of industry.

In the official statistics, it is only *new* construction which is part of the investment goods producing branch. Repair and maintenance construction is seen as the production of consumption goods. The differences between demand for investment and consumer goods are reflected in the data, which shows greater instability in new than in repair and maintenance construction demand.

The reasons for this greater short-run variability of investment demand are complex. However, to some extent the relatively greater short run variability of investment demand is caused by the fact that investment decisions are based upon expectations of a more distant future than are consumption decisions. Such distant expectations are inherently more uncertain, more unstable, and more likely to be influenced by such irrational intangibles as the level of investors' confidence in the future. Another reason is that the motive for corporate investment is the expected profit return, and expected rates of profit are one of the more unstable and unpredictable of economic variables. Moreover, most corporate investment is actually financed out of firms' retained profits from their recent actual trading activity, and actual or *ex post* levels of profit also fluctuate widely. Finally, the accelerator effect in theory also accounts for the variability of investment demand.

Investment decisions involve both current and future production. Investment in stocks of finished but unsold goods, or of inputs and work in progress, is a function of current production and sales. Investment in fixed capital is a function of expected future sales. These two kinds of investment are measured in the *Blue Book*.

Current production is made up of production for sale and additional production for stock. During periods of business optimism and economic confidence and growth, the value of output, and the value of the assets used to make that output, rise. Production increases. If, however, current sales patterns indicate that a slow down is taking place, firms respond by cutting production of goods for stock. Indeed, production can be less than sales if the firm decides to de-stock. As a result, current production may even begin to decline although the rate of sales growth, still taking place, has only begun to slow down.

Now, as far as the construction industry is concerned, it is only speculative developers who engage in production for stock. They hold variable stocks of finished but unsold output, in contrast to contractors, who by definition only produce to order. In Tables 9.1 and 9.2 partly taken from the *Blue Book*, we show a time series of changes in book value of stocks by industry, and the value of the physical increase in stocks by industry. The construction industry stock changes series are large in average mag-

Table 9.1 Investment in fixed and circulating capital, at current market prices, 1985–94, £million.

Year	GDFCF, all fixed assets		Change in book value of stocks		All investment in fixed and stock capital		GFCF as % of all investment	
	All inds.[1]	Constr. Ind.	All inds.	Constr. Ind.	All inds.	Constr. Ind.	All inds.	Constr. Ind.
1985	48,499	626	3,559	1,268	52,058	1,894	93.2	33.1
1986	50,892	609	2,517	1,250	53,409	1,859	95.3	32.8
1987	58,803	763	5,955	2,030	64,758	2,793	90.8	27.3
1988	70,603	1,142	10,708	3,304	81,311	4,446	86.8	25.7
1989	82,455	1,111	9,738	2,757	92,193	3,868	89.4	28.7
1990	86,138	965	4,331	1,298	90,469	2,263	95.2	42.6
1991	79,246	585	−2,917	−156	76,329	429	103.8	136.4
1992	74,908	563	−159	−412	74,749	151	100.2	372.8
1993	75,063	650	2,679	29	77,742	679	96.6	95.7
1994	79,123	727	7,851	766	86,974	1,493	90.1	48.7
1995	86,481	821	9,509	817	95,990	1,638	90.1	50.1
1996	92,085	1165	3,890	44	95,975	1,209	96.0	96.4
All years	884,296	9,727			941,957	22,722	93.9	42.8

Note:
1. All industries represents the whole economy less ownership of dwellings.
Source: National Accounts, (1997 edn), Tables 13.8 and 15.1 (London: HMSO).

Table 9.2 Total investment and its composition, by industrial sector, at current market prices, 1985–96, £ million

Year	Primary sector (agriculture; mining)				Manufacturing sector			
	GFCF in all fixed assets	Change in book value of stocks	New buildings and works[1]	GFCF in structures as % of all GFCF	GFCF in all fixed assets	Change in book value of stocks	New buildings and works	GFCF in structures as % of all GFCF
1985	5,149	-423	2,544	49.41	10,283	648	1,207	11.74
1986	4,832	-505	2,179	45.10	10,105	-48	1,203	11.90
1987	4,491	-131	2,171	48.34	11,040	1,713	1,356	12.28
1988	4,956	-38	2,485	50.14	12,415	3,256	1,578	12.71
1989	5,540	265	2,791	50.38	14,248	2,483	1,960	13.76
1990	6,068	33	3,408	56.16	14,227	859	1,907	13.40
1991	7,021	146	4,672	66.54	13,183	-2,688	1,811	13.74
1992	6,813	91	4,522	66.37	12,433	-135	1,454	11.69
1993	6,071	-349	3,965	65.31	12,410	-241	1,315	10.60
1994	4,742	22	3,216	67.82	13,534	3,377	1,517	11.21
1995	5,402	-14	3,591	66.48	15,775	5,152	2,064	13.08
1996	5,749	-415	3,767	65.52	15,388	-17	1,989	12.93
All years	66,834		39,311	58.82	155,041		19,361	12.49

Table 9.2 Continued

Year	Utilities sector (gas, water, electricity)				Trade sector (retail, wholesale, hotels and rents)			
	GFCF in all fixed assets	Change in book value of stocks	New buildings and works	GFCF in structures as % of all GFCF	GFCF in all fixed assets	Change in book value of stocks	New buildings and works	GFCF in structures as % of all GFCF
1985	2,660	396	1,132	42.56	5,739	1,108	1,702	29.66
1986	2,792	62	1,255	44.95	6,269	2,156	2,098	33.47
1987	2,798	-112	1,154	41.24	7,687	2,777	3,125	40.65
1988	3,119	228	1,226	39.31	9,456	3,860	3,726	39.40
1989	3,943	113	1,422	36.06	9,468	3,580	3,932	41.53
1990	4,742	-67	1,786	37.66	9,136	1,675	3,826	41.88
1991	5,608	11	2,412	43.01	8,352	-449	3,524	42.19
1992	6,365	14	2,753	43.25	8,225	1,140	3,456	42.02
1993	5,910	-214	2,449	41.44	7,936	3,395	2,851	35.92
1994	5,221	-619	2,108	40.38	8,263	3,930	2,713	32.83
1995	5,085	-170	1,664	32.72	10,744	3,615	3,787	35.25
1996	4,567	15	1,385	30.33	11,352	2,303	4,346	38.25
All years	52,810		20,746	39.28	102,367		39,086	38.08

Table 9.2   Continued

	Construction				All sectors			
Year	GFCF in all fixed assets	Change in book value of stocks	New buildings and works	GFCF in structures as % of all GFCF	GFCF in all fixed assets	Change in book value of stocks	New structures and works	GFCF in structures as % of all GFCF
1985	626	1,268	65	10.38	60,718	3,559	27,437	45.19
1986	609	1,250	45	7.39	65,032	2,517	30,654	47.14
1987	736	2,030	103	13.99	75,158	5,955	36,229	48.20
1988	1,142	3,304	127	11.12	91,530	10,708	45,721	49.95
1989	1,111	2,757	138	12.42	105,443	9,738	54,356	51.55
1990	965	1,298	176	18.24	107,577	4,331	56,294	52.33
1991	585	−156	112	19.15	97,747	−2,917	50,260	51.42
1992	563	−412	120	21.31	93,642	−159	48,132	51.40
1993	650	29	39	6.00	94,293	2,679	46,253	49.05
1994	727	766	25	3.44	100,252	7,851	47,635	47.52
1995	821	815	88	10.72	108,736	9,509	51,097	46.99
1996	1,165	44	208	17.85	114,623	3,890	54,317	47.39
All years	9,700		1,246	12.85	1,114,751		548,385	49.19

Note:
1. Works include oil rigs and the like. Thus, not all GFCF on new buildings and works constitutes demand for construction industry, as defined in the SIC.

Source: Blue Book, 1996, 1997, Tables 13.8 and 15.1.

nitude relative to other industries and relative to construction industry fixed capital formation. They also fluctuate widely.

During the upswing between 1985 and 1990, the construction industry accounted for a high proportion of the total increase in book value of stocks in the whole economy. This amounted to £11,907 million out of a total increase in the whole economy of £36,808 million. At the end of 1990 the construction industry held stocks with a total book value of £13,730 million. This can be compared with total book value of stocks in all industries combined of £124,216 million. Then, during the downswing in 1991 and 1992, the book value of construction industry stocks fell by £568 million compared to a fall in all-industry stocks of £3,076 million.

Moreover, whereas for the whole economy increases in the value of stocks are always small when compared to gross domestic fixed capital formation, this is not so for the construction industry (see Table 9.1). This fact has one other significance altogether, which we explore in the companion volume when we consider the need to finance investment for growth by construction firms. It means they mainly need short term finance.

Here our concerns are with the demand for construction and hence with all investment expenditure by promoters of construction projects. These promoters as we have seen divide into those external to the construction industry on the one hand and speculative builders on the other.

Table 9.2 shows the proportion of investment in fixed capital which is invested in buildings and works for different sectors of the economy. The primary sector requires a smaller proportion to be invested in plant and equipment than manufacturing. The figures show a lower proportion of buildings and works in the fixed investment of manufacturing than in other sectors of the economy. At the same time, the data appear to show that the primary sector spends a higher proportion of its total investment on new structures than do other sectors.

Where there is an external promoter or client, the statistical trace left by an investment decision has this sequence:

- first, a temporary increase in orders to the construction industry;
- next, a temporary increase in the total physical stocks, and their book value, held by the construction industry, representing increased work in progress. This figure is absolutely quite small, because once monthly instalments have been valued, they no longer appear as

work in progress in the accounts of the construction industry. Thus, it is the increase in monthly rather than annual output that appears;

- next, an increase in gross and net output of the construction industry including value added by the output based method of measurement; and
- gross fixed capital formation (GFCF) on new buildings and works reported for the industry to which the promoter or client belongs;
- finally, a permanent increase in the gross capital stock (GCS) of the industry of the client.

Where on the other hand the promoter is a speculative construction industry firm the statistical trace is as follows:

- first, an increase in stocks held by the construction industry, representing increased stocks of inputs and then work in progress. This is absolutely large as it represents the whole cost of inputs including land purchased throughout the development process of the project;
- also an increase in *orders* to the construction industry (though these are not, in truth, external orders);
- then, an increase in output and value added of the construction industry reported as and when the project is sold to an external buyer;
- at the same time GFCF reported for the industry to which that buyer belongs, including the ownership of dwellings industry;
- finally, an increase in the GCS of that industry.

In principle GFCF can be measured *ex ante,* as soon as the decision to invest has been made. Perhaps in practice this may be measured when contracts for supply are signed. Alternatively, GFCF can be measured *ex post,* when the investment expenditure has been made and the investment goods delivered. In practice, as can be seen in Table 9.2, the *Blue Book* reports GFCF *ex post,* so that it corresponds in time to output rather than orders.

At times, as speculative builders' sales increase to a level higher than production, stocks will begin to be depleted. If firms believe the increase in sales is very temporary, they may allow stock depletion to act as a buffer, protecting production volumes from short term changes in sales volumes. But when they become confident that the increase in sales is more permanent, and simultaneously worried that they may run out of stocks from which to meet demand, there will be a deliberate surge in production to raise stocks. The level of production then exceeds the level of sales. Obtaining credit by firms to expand by holding larger stocks becomes easier as their growing asset base (which acts as collateral for loans) combined with banking confidence make it easier for firms to borrow to invest. This was experienced in the mid-to late 1980s, although over-borrowing and over-confidence had been a characteristic of previous booms in economic activity prior to recessions.

In conventional theory concerning the business cycle and circulating capital investment, firms are assumed to produce ahead of making sales, with output first being stockpiled, and then sales being met from stocks. This is by definition not true of contract construction. Holding of unsold stocks and stock adjustment are nevertheless very important in construction materials production and in speculative development of property, including housebuilding. In housebuilding land stocks and work in progress held by developers fluctuate greatly, in volume and in value. In contracting, though, the stocks of circulating capital goods held by firms have mostly been pre-sold. They remain as circulating capital goods, rather than as cash, simply because though sold they have not yet been paid for. In contracting firms, such stocks will only fluctuate in proportion to sales, rather than disproportionately as in the conventional stock-adjustment model of the business cycle.

The other form of investment concerns the preparations necessary for production in some future period. These investment decisions concern the future production capacity of the firm or the industry and involve a longer term horizon than the current sales period. Capital investment decisions depend on the confidence of decision makers that a market for the future output will exist. To undertake production in the future, investment may be needed in research and development, new plant and machinery and of course new buildings, such as factories and offices, as well as the spatial infrastructure.

As we have noted, much of the demand for construction is a derived demand. Owner-occupier firms do not want buildings for the sake of owning the buildings but need them for what can be profitably produced within them. Production will only be profitable if there is an expected effective demand for future output. The demand for buildings depends directly on the expected demand for final products or demand for services.

However, not all demand for construction from speculative property-holders can properly be called derived demand, as it is based on expected increases in property prices, resulting from anticipated shortages of property supply. Not only may this kind of speculative demand be stimulated by supply restrictions as much as by demand increases, but, unlike derived demand it is not necessarily linked to forecasts of increased need for space in any one particular industry or activity.

Speculative developers' expenditure on new buildings is not envisaged by them as being a fixed capital investment. Rather, it is an investment in working capital to finance production. The fixed capital investment decision is that made by the eventual buyer. Thus speculative developers' investments in working capital (in the form of stocks of buildings) are attempts to *anticipate* the fixed investment demand by others for fixed capital assets. Nevertheless, fixed capital formation has in a real sense occurred prior to sale and use of a building – the asset has been produced, and resources diverted to its production.

We now need to introduce two key concepts used in economic theories of cyclical behaviour of the economy: the multiplier and the accelerator.

The multiplier

The simplest approach to the multiplier begins by assuming no foreign trade and no government intervention. Savings depend on income. In Figure 9.1, savings and investment are measured on the vertical axis, income along the horizontal.

Hence:

$$S = f(Y) \tag{9.1}$$

where

$$S = \text{savings}$$

and

$$Y = \text{income.}$$

At Y_1,

$$S = I \tag{9.2}$$

where

$$I = \text{investment.}$$

Assume Y_1 is an equilibrium value of Y, and point A is the equilibrium combination of the income co-ordinate and S/I coordinate. National income depends upon the level of investment. Now suppose that there is an autonomous change in the level of investment demand, to I_2 (here, 'autonomous' indicates that the change is caused by reasons other than a change in national income itself). A rise in I eventually increases income, aggregate demand and savings until a new equilibrium is reached at B. If at point B, at income level Y_2, but not at any lower level of Y, the existing stock of capital and labour is fully utilised, then Y_2 is called the full-employment income level. If B is also an S/I equilibrium, because the increase in Y from Y_1 to Y_2 has caused an increase in S equal to the increase in I, it is the special case known as the full-employment equilibrium. If *animal spirits* lead firms to attempt an even higher level of I than corresponds to point B, they will find that output, and thus real income, cannot be increased. Either investment will increase but only at the expense of a reduction in consumption, or increased demand for investment goods will only lead to increased prices for these. In neither case will real income increase, and thus neither will savings. A higher level of I is unobtainable. However, it is important to be clear that the ultimate reason for this is not the lack of sufficient savings to finance the investment, but the lack of under-utilised productive capacity, which prevents the increase in real output (and hence incomes) that could otherwise generate extra savings.

The multiplier effect demonstrates the impact a change in investment can have on the levels of income and employment in an economy. The investment multiplier is the ratio of the change in national income to the change in investment which brought it about.

Let

$$m = \Delta Y/\Delta I \qquad (9.3)$$

where

$$m = \text{the multiplier}$$
$$\Delta Y = \text{change in national income}$$
$$\text{and } \Delta I = \text{change in investment.}$$

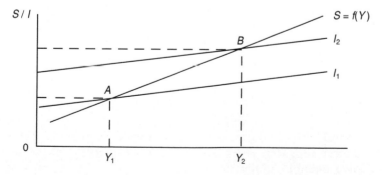

Figure 9.1   Effect of increase of investment on national income and savings

The multiplier effect on national income is therefore the change in investment multiplied by the multiplier. The multiplier effect means that any change in investment will have a greater effect on the economy than the change in investment which causes it. For instance, if the multiplier is equal to 4, then an increase in investment of £300 million will raise the level of national income by £1 200 million. The multiplier magnifies the effect of an investment by increasing the value of aggregate demand by an amount greater than the value of the original investment. This short-run effect is quite independent of the long term contribution which investment will make to capacity, and would indeed still occur even if the investment plant or equipment were never used, so long as investment goods are produced and bought. It works through putting more people to work in the investment goods branch of production, and therefore putting more wage-income in their pockets, and more profit income in the hands of capitalists making investment goods, part of which they then spend on consumption goods. It only operates if the economy starts from a position of spare capacity and unemployment. Otherwise, extra investment demand would merely force up prices of resource inputs and labour, and at best redistribute some of a given total of productive resources away from making consumer goods and towards making investment goods.

In the simplified model, the derivation of the value of the multiplier is linked to consumption and saving, as follows:

let

$$m = 1/(1 - mpc) \tag{9.4}$$

where

m = the value of the multiplier
mpc = marginal propensity to spend additional income on consumption.

Since additional income must either be spent or saved,

$$(1 - mpc) = mps \tag{9.5}$$

where

mps = marginal propensity to save.

Substituting (9.5) in (9.4), the multiplier can also be defined as:

$$m = 1/mps \tag{9.6}$$

In the example above, if the $mps = 0.25$ (25 per cent of any additional income received will be saved, and 75 per cent spent), then the value of the investment multiplier must be:

$$m = 1/0.25 = 4 \tag{9.7}$$

Of the initial extra £300 million of income distributed as wages and profits in the capital goods-making sector as a result of the £300 million increase in investment, 75 per cent or £225 million must in turn have been spent, in what is known as the 'first round' increase, on additional consumption. This will have increased output of the consumption goods industries, and their income, by £225m, distributed as wages and profits. Of this £225m, 75 per cent or £169m will in turn have been spent. And so on. Thus we have a mathematical series, in which each term has 3/4 the value of the preceding term. The sum of the value of this series is four times the value of the initial term.

Note that, since the mps = 0.25, the increase in savings resulting from the multiplied increase in income will be one-quarter of £1200 million, or £300 million – just equal to the increase in investment. Thus S and I remain equal, but at a higher level. If the mps had been higher than 0.25, the final multiplied increase in income resulting from the investment would have been less, but the new S would still have been equal to the new I.

The accelerator

Let us now turn to the relationship between the demand for consumer goods and services and the demand for the built environment. Here we must distinguish between final consumer goods and the means of production. Final consumer goods are purchased by consumers for their own use. Non-durable goods such as food can only be used once, whereas durable goods can be used many times over a period of time. Durable consumer goods are generally not resold except at a lower second hand price, which reflects the fact that they lose some of their value due to being worn out to a greater or lesser extent through use.

Final consumer goods are bought as an end in themselves to satisfy a consumer's requirement. In contrast, the means of production are required not for their own sake but for the value of the commodities they can produce. Hence investment in the means of production is based on the expected sales of the goods to be produced by the additional plant and machinery in additional buildings. Similarly the justification for investment in public sector infrastructure is the improvement to production and distribution, capacity and output, which new roads, public transport links and the like facilitate.

A small change in the rate of change in demand for consumer goods and services may cause a large change in the demand for plant and equipment and buildings, used to produce those goods and services.

This is demonstrated by the accelerator principle, which is based on the amount of fixed capital stock required for a given level of output. This ratio is known as the capital to output ratio and is the key to understanding the accelerator principle.

For instance, if a firm produces an annual output of £100,000 and requires a stock of £300,000 of fixed capital for buildings, plant and machinery in order to do so, its average capital to output ratio is 3. Assume a constant capital to output ratio so that the marginal ratio is equal to the average ratio. A planned rise of, say, 10 per cent in output would therefore require an increase of 10 per cent in the capital stock. In this example, a £10,000 increase in the flow of annual output would require a £30,000 once-and-for-all increase in the capital stock.

Every year some plant and machinery and buildings need to be replaced. They may be worn out or the cost of maintenance may have become greater than the cost of replacement. Let us assume that a firm uses 10 machines bought at different times, each with a planned life of 5 years. At the end of 5 years the machines wear out and must be replaced. Let us also assume that there is no change in the level of output. Even without a change in the level of output in a given year, 20 per cent of the plant or 2 machines on average would still need to be replaced annually just to maintain production levels. This then will be the *base* level of demand for machines. If demand for the output produced by the firm were to rise by, say, 20 per cent, then the firm would need to increase the number of its machines from 10 to 12 to meet the extra demand with extra production. The amount of total investment in plant would rise by 100 per cent, because the firm would not only need to replace 2 worn out machines, but it would also need 2 extra machines.

The demand for the means of production includes the demand for buildings and the built environment. It is the relationship between consumer demand and the demand for investment in the means of production which is relevant for our purposes. If the amount of increase in consumer demand between years stays constant, this is consistent with a steady positive *level* of net investment. But, because of the interplay of accelerator and (lagged) multiplier, the economy will tend to fluctuate around this rate of growth. Investment grows from year to year only if output grows at an *accelerating* rate.

The lifespan of means of production makes an important difference to the responsiveness of investment demand to changes in consumer demand. The longer the lifespan the greater the impact of changes in consumer demand on demand for means of production. This is again particularly important in construction because of the very long life of buildings and infrastructure and partly explains the great variation in demand which often faces firms involved in the production of the built environment.

If the lifespan of fixed capital goods is short, changes in consumption demand will have relatively less impact on investment demand than if life spans are long. This can be demonstrated as follows:

Let

I = annual total investment demand (gross investment)

R = annual replacement investment demand

K = capital stock

ΔK = change in capital stock (net investment)

L = lifespan of items of fixed capital goods

then

$$I - R + \Delta K \qquad (9.8)$$

or

$$\Delta K = I - R. \qquad (9.9)$$

In a steady state economy, with no growth or decline in demand or output, there would be no need to change K.
Hence, if

$$\Delta K = 0 \qquad (9.10)$$

and

$$L = 5 \text{ years,} \qquad (9.11)$$

then

$$R = 20 \text{ per cent of } K \qquad (9.12)$$

and

$$R = I. \qquad (9.13)$$

At a rate of replacement of 20 per cent per annum all plant would be replaced after 5 years in time to meet its lifespan without loss or gain in the amount of plant.
Thus, if

$$K = 300 \text{ (measured in units of investment demand)} \qquad (9.14)$$

then

$$R = 60 \qquad (9.15)$$

and

$$I = 60 \qquad (9.16)$$

But, if

$$L = 20 \text{ years} \qquad (9.17)$$

then

$$R = 15 \qquad (9.18)$$

and

$$I = 15 \qquad (9.19)$$

In (9.18), only 15 units of replacement investment would be required annually to maintain K at 300 instead of 60 units as in (9.15). The shorter the lifespan the more annual investment is required simply to replace and maintain the existing capital stock.

Now assume

$$O = D \qquad (9.20)$$

where

$$O = \text{output}$$

and

$$D = \text{consumer demand (measured in units of output).}$$

Let

$$D = 100 \qquad (9.21)$$

From (9.20),

$$O = 100. \qquad (9.22)$$

Let

$$K{:}O = \text{capital: output ratio}$$

If

$$K{:}O = 3{:}1 \qquad (9.23)$$

and

$$O = 100 \qquad (9.24)$$

then

$$K = 300. \qquad (9.25)$$

Now, let

$$\Delta D = \text{change in consumer demand}$$

If

$$\Delta D = 10 \text{ per cent of } 100 = 10 \qquad (9.26)$$

then

$$\Delta K = 10 \text{ per cent of } 300 = 30. \qquad (9.27)$$

Now, if

$$L = 5 \text{ years}, \qquad (9.11)$$

substituting (9.15) and (9.27) in (9.8):

then

$$R + \Delta K = 60 + 30 = 90 \qquad (9.28)$$

But, if

$$L = 20 \text{ years}, \qquad (9.17)$$

substituting (9.18) and (9.27) in (9.8):
then

$$R + \Delta K = 15 + 30 = 45. \tag{9.29}$$

From these equations we can see that in response to the 10 per cent rise in demand and output, if $L = 5$ years, annual total investment rises from 60 units (9.16) to 90 units (9.28), an increase of 50 per cent. Whereas, if $L = 20$ years, the increase in investment is 200 per cent, up from 15 (9.19) to 45 units (9.29).

Put another way, the longer the lifespan of fixed capital goods, the longer it will take for any given percentage increase in the annual output of capital goods to achieve the desired increase in the capital stock, caused by the increase in consumption demand. If $L = 5$, 'steady state' output of capital goods is equivalent to 20 per cent of the fixed capital stock. Doubling output of investment goods raises K by 20 per cent per year. It would then take 5 years to double K. Hence, time taken to double K, assuming output of investment goods can only increase by a certain percentage over its base level, depends on the replacement life of K. If $L = 20$ years, 5 per cent of K will be replaced each year. Doubling investment raises K by only 5 per cent and it would take 20 years to double K.

The size of the accelerator coefficient is equal to the incremental ratio of capital to output. As the capital value of buildings and plant used in production is greater than the annual value of output, then, assuming the capital to output ratio is constant, an increase in output leads to larger increases in capital employed.

Capital output ratio $= \dfrac{K}{O}$ (9.30)

where

K = capital required

and

O = annual output

$$v = \frac{\Delta K}{\Delta O} = \frac{K_{t+1} - K_t}{O_{t+1} - O_t} \tag{9.31}$$

where,

v = accelerator coefficient
ΔK = change in capital required
ΔO = change in annual output
K_t = capital required in period t
K_{t+1} = capital required in period $t+1$

$$O_t \ = \ \text{annual output in period } t$$
$$O_{t+1} \ = \ \text{annual output in period } t+1$$
Transposing from (9.31):
$$v.\Delta O \ = \ \Delta K \tag{9.32}$$

Equation (9.32) therefore states that a change in the value of planned annual output multiplied by the accelerator co-efficient provides the change in capital required to meet expected output.

An example may help to clarify the accelerator principle. In any given period, the capital and output figures are in a common measure called units. Assume a constant capital to output ratio of say, 3:1. If planned output rises from 10 units to 12, then capital required must rise from 30 units to 36. It follows that an increase of 2 units of output requires 6 extra units of capital, giving an accelerator co-efficient of 3. This is illustrated in Table 9.3 and occurs in the third period.

As stated, this example assumes a constant capital output ratio, but this assumption is unrealistic, especially when applied to the built environment. We have seen that firms can adopt a variety of strategies to intensify their use of existing buildings. Often in firms there is excess capacity or slack, which can be utilised to increase production without necessarily increasing capital employed. This has the effect of reducing the *incremental* capital to output ratio, though not the average ratio.

In any case new technology invariably alters the capital to output ratio as more efficient capital replaces worn out assets. Either less capital is required for a given level of output or more capital is required but in the process is used as a cheap substitute for expensive labour. Indeed if this were not the case, firms would not replace their equipment as quickly as they do. Moreover a rise in consumer demand may be met out of stocks of finished goods without any need to increase output let alone invest in extra means of production. Indeed unless

Table 9.3   Simple model of accelerator effect

Period	Capital	Output	Capital:output ratio	Change in output	Accelerator coefficient	Change in capital
1	30	10	3:1	0		0
2	30	10	3:1	0		0
3	36	12	3:1	2	3	6

firms are confident that expansion in demand is likely to be long term, it is unlikely that increases in demand would call forth extra production, still less investment. Extending lead times, creating customer waiting lists and price increases are alternative responses.

Table 9.4 shows a more elaborate analysis of the accelerator applied to the built environment. In this model, existing built stock shows the quantity at the beginning of each period. Assume all exogenous demand is consumer demand and that output, other than of investment goods, in any year is equal to that demand. Built stock required is the quantity of built stock needed to maintain a constant capital output ratio of 3:1. Retirements are assumed to be at the rate of 2 per cent per annum (i.e. the average stock lifespan is 50 years).

In particular the two points to note from Table 9.4 are firstly the greater variability of investment demand compared to consumer demand and secondly the non-concurrence of consumer and investment demand even when time lags in production are ignored. While consumer demand varies by ±1 per cent in the model, investment demand rises by 50 per cent in year three from 6 to 9 units only to fall again by 33 per cent from 9 to 6 units in the following period and then to fall by a further 50 per cent from 6 to 3 units in period 6.

In reality, it should be noted that retirement and replacement of built stock can often be postponed by firms cautious about the future. Delays in investing in the built environment in response to increases in demand would distort the accelerator co-efficient as an accurate predictor of the behaviour of the economy following increases in consumer demand. Increases in demand may also emanate from the public sector as part of government policy aimed at stimulating the economy. However, it is the response of firms to increases in demand based on their expectations and confidence as well as spare capacity and the level of holdings of stocks which determine the level of corporate sector investment.

Sherman (1991) notes it is relatively straightforward to introduce time lags into the accelerator principle. Thus, let investment in one period be determined by the change in demand for output in the previous period. In algebraic terms:

$$N_t = v\,(O_{t-1} - O_{t-2}) \qquad (9.33)$$

where

$$N = \Delta K$$
$$N_t = \text{net investment in period } t$$
$$O_{t-1} = \text{annual output in period } t{-}1$$
$$O_{t-2} = \text{annual output in period } t{-}2.$$

Table 9.4 Accelerator model ignoring time lags, expectations and spare capacity

Period	1	2	3	4	5	6	7	8
Consumer demand = output of consumption goods	100	100	101	101	101	100	100	99
Change in consumer demand	0	0	1	0	0	−1	0	−1
Existing built stock	300	300	300	303	303	303	300	300
Built stock required	300	300	303	303	303	300	300	297
Change in built stock	0	0	3	0	0	−3	0	−3
Retirements (approx)	6	6	6	6	6	6	6	6
Replacements	6	6	6	6	6	3	6	3
Net investment	0	0	3	0	0	0	0	0
Total investment demand	6	6	9	6	6	3	6	3
Change in total investment	0	0	3	−3	0	−3	3	−3

The actual size of the accelerator co-efficient is largely determined by the behaviour of firms in response to changes in demand. Output is not linked *technologically* to demand, but to the strategies which firms adopt, their plans and *ad hoc* decisions. Their investment decisions are taken on the basis of their assessments of past, current and future output, as well as their competitive performance and market expectations. Moreover, the decision to invest not only is concerned with changes in demand but also considers anticipated costs of production such as labour, materials and interest rates, which would all affect the incremental capital to output ratio.

Finally, the decision to invest depends on the level of existing stock and spare capacity. This is particularly relevant for decisions to invest in buildings. Several strategies can be suggested which would allow firms to delay or avoid investment in buildings in response to an increase in demand. This implies that the relationship between demand and investment is far more variable than suggested by the accelerator principle. For instance, before firms need to invest in new buildings, it may be possible for them to extend the length of the working day by working overtime. Alternatively, rearranging the plant and equipment within a building may improve the use of space without requiring additional floor space.

Change-induced demand for buildings and works

Firms compete by continuously adopting new techniques and technologies. This process of innovation constantly requires changes to the built environment in the form of new building types. Moreover, changes in demographic structures and the migration of populations from one region to another call for new building and infrastructure to accommodate the changes. As a result of these and many other changes the stock of buildings and works is in constant need of adaptation, replacement and modernisation.

These exogenous changes to the required built environment further complicate the accelerator model which relates only to corporate sector investment demand as a function of changes in the level of consumer demand. It is therefore clear that total investment demand for construction services depends on consumer income, public sector demand, and exogenous economic, social and political factors. Time lags will also occur as firms postpone investment decisions, and the lead-in times for planning projects and then carrying out construction mean that often there is considerable delay before investment projects come on stream. These factors can, to some extent, be included in assumptions made in the model of the accelerator shown in Table 9.4, for instance by the addition of a row for public demand and by showing the timing of capital expenditure in the appropriate cell in the table.

Moreover, the insertion of speculative property developers and property landlords into the investment process modifies it fundamentally. For much of the built stock it is these developers who actually take the investment decisions. These are of course still based in part upon reactions to, and forecasts of, change in demand for the output of building users (and hence the latter's demand for built space). But, necessarily, developers cannot have such close or advance knowledge of these changes, and deal in the generality of demand rather than specific demand facing individual firms. Moreover, competition between developers increases the likelihood that some will fail to find users (tenants) for their projects.

Concluding remarks

This chapter has shown that construction demand is a key element in the structure of investment demand. Gross fixed capital investment in buildings and works varies from just over 10 per cent of all investment in manufacturing to around 65 per cent in agriculture and mining. Investment in plant and machinery and vehicles make up the rest of GDFCF, which is itself normally over 90 per cent of all investment. From Table 9.2 it can be seen that in absolute terms, more investment in structures took place in the financial and business services sector than in any other sector.

As construction demand is such an important component of investment demand, it plays a major role in stimulating economic activity in the rest of the economy. This is because demand for construction arises out of business decisions which depend on confidence and expectations. These decisions to act in any one period then bring about large injections of money expenditures and incomes into the economy. If there were no spare capacity or unemployed resources, the additional investment expenditure would be inflationary. In principle, given spare capacity in the economy, investment in structures will have a multiplier effect, and generation of demand for construction by growth in the rest of the economy can be seen in the accelerator principle.

Changes in the level of economic activity depend on investment and investment depends on expectations and confidence concerning the future. An economic model of the relationship between investment and the level of activity has been developed by several economists, including Samuelson and Hicks. Their model of the business cycle is based on the interaction of the multiplier effect and the accelerator principle.

Put simply investment has two effects on economic activity. First it increases incomes as employment is created. This is the multiplier effect. Second, it increases capacity, and it is the accelerator principle which describes the relationship between investment, capacity and output. Eventually, if the increase in capacity is greater than the increase in the ability of the economy to absorb the extra production, the result is stagnation or depression with unemployment and under-utilised plant and equipment.

We are now in a position to apply the multiplier effect and the accelerator principle to the trade cycle. What can be observed with the accelerator is the inherent instability in any capitalist economic system. Investment in plant and machinery as well as buildings is timed differently and yet is dependent on the cycle of demand in the consumer goods and services sector.

As we have noted, replacement investment is needed to maintain the building stock at its current level. Total investment less replacement investment is equal to net investment. Net investment is therefore the amount of investment over and above the amount of investment needed to replace worn out stock. It is this distinction between replacement investment and net investment that is important in applying the accelerator principle to the business trade cycle.

Now, whilst the accelerator effect is an important idea, two of its assumptions are quite unrealistic: namely, that investment goods have a predetermined fixed lifespan, and that more output can normally only be produced by increasing the fixed capital stock by a technically determined amount.

Actual empirical studies show that, at least in the medium term, capital-output ratios have variable rather than fixed values. This is because many industries actually operate most of the time with a certain amount of spare capacity, and can thus, if demand permits, extract more output from their given capital stock (Brenner, 1998). Moreover, to some degree other inputs to production can substitute for any one type of capital good for which there is no excess stock. In any case, *capacity* is a more socially fluid concept than the technically determined notion suggested by the accelerator theory. For example, it is usually possible to increase the output of plant by working longer hours or to increase the capacity of buildings by rearranging work spaces to fit in more people or machinery.

Furthermore, and this applies especially to buildings and infrastructure, capital goods do not disintegrate when their expected useful life has expired. When demand is high, old fixed capital can be pressed into extended life. There will be a cost, in higher maintenance and operating charges. However, owners of capital goods may find this worth paying, when output prices are high and they can sell all that they can produce. On the other hand, unexpected technological change, resulting in competition from superior, new fixed capital embodying new technology, may result in the premature demise, obsolescence, and scrapping of fixed capital well before the end of its expected life.

Time lags are also important. It takes time for additional demand for consumption goods to be translated into additional output. More time will elapse before all the resulting incomes are fully distributed as wages and profits. Further delay is possible before these incomes are then spent. All of this will reduce the size of the multiplier effect in the first year following the injection of investment funds.

In the meantime, as a result of these lags, a *savings gap* may emerge. The need to finance the extra investment arises with relatively little lag, but the source of that extra funding is extra savings. These additional savings arise out of increased incomes but because of time lags incomes do not rise sufficiently to produce a sufficient increase in savings in time to meet demand for increased investment. This may push up interest rates. Likewise, if demand increases faster than industries can respond, temporary shortages may emerge, pushing up prices. If these price rises threaten to turn into a generalised surge in inflation, the initial increase in investment demand may be reversed, particularly by government intervention.

Moreover, modern economies are not closed to the rest of the world. The investment multiplier's effect on national income is greatly weakened in an open economy, where a high proportion of any extra household income will be spent on imported consumption goods. Moreover, apart from the international trade in goods, there are significant international flows of lending and borrowing. As a result, national investment and savings need no longer be identical. For example, much of the 1980s investment boom in the UK property and housing market was financed with Japanese and other countries' savings, lent through the City of London to UK developers.

At first sight this looks as if it removes a potential savings constraint on the ability of an economy to sustain an investment-led boom. However, the availability of such saving flows depends *inter alia* on international relativities in interest rates, upon expected relative movements in exchange rates, and upon the underlying long term willingness of international financial institutions to increase the size of their portfolios of financial assets held in the UK. These constraints on international lending mean that in the longer run, there is a need to maintain incomes and savings at levels which inspire international confidence in the ability of the UK economy to repay any loans, whilst simultaneously maintaining relatively high interest rates and exchange rates. The greater that long term confidence, the lower the level of interest rates needed to attract a given inflow of funds.

10
Business Cycles and Construction

Introduction

Variations in aggregate construction demand follow a cyclical pattern, as does demand in the economy as a whole. We develop a business cycle-based explanation of the cycles in construction demand, and advance a profits-based model of the endogenous operation of the business cycle.

We survey different methods of measuring trends and cycles in raw time series data, and apply a method of describing cycles developed by the US National Bureau of Economic Research. We compare trends and cycles in GDP with trends and cycles in construction output.

A profit-led model of private sector investment demand and of business cycles

Because a large part of construction demand, is dependent on investment demand in the industrial and commercial sectors of the economy, we are concerned with the total profit of the UK business sector. To look at factors affecting this, we proceed as if that sector consisted of one giant firm with many departments ('UK plc'). We ignore changes in prices at which one part of that 'firm' trades with other parts, and concentrate only upon its *external* purchases of labour and of imports and its sales to households, government and exports. Our theory assumes that total construction demand by the UK business sector will depend upon its profits and its expectations of profits.

We need to distinguish the terms *profit margin, mass of profit, and rate of profit*. Profit margin is the profit per £ of sales or turnover and is best expressed as the mass of profit as a percentage of the value of sales. Mass of profit is the total profit in £s, the profit margin multiplied by the value of sales. Rate of profit is the mass of profit divided by the value of capital owned. The rate of profit is also known as 'profitability' the return on capital employed (ROCE) and is the multiple of two ratios. The first of these ratios is the profit margin and the other is the ratio of the value of sales to the value of capital employed. Thus:

$$ROCE \;=\; (P/T)(T/K) \tag{10.1}$$
$$=\; P/K \tag{10.2}$$

where

P = Profit
T = Sales or turnover
K = Capital employed

The profit margin per unit of output is, of course, selling price per unit minus cost per unit.

Costs fall under two main headings, namely fixed capital costs and direct costs. Fixed capital costs depend for their mass on the amount of past investment, the size of the capital stock, and the rate at which the stock is depreciating or becoming obsolete. Expressed per unit of sales, fixed capital costs per unit tend to rise in recessions, because capital stock is not fully utilised, and does not fall as fast as output or sales. Indeed total fixed capital costs may peak during the crisis and early recession, as investment commitments planned during the boom are delivered and have to be paid for. Whereas, in early recovery, capital costs per unit may fall, as spare capacity comes back into use, and output can rise without requiring any increase in fixed capital stock. The profit margin is sometimes expressed as the gross margin of selling price over prime cost, and sometimes the net margin over total cost, including capital cost, per unit.

Unlike capital costs per unit, direct or prime costs per unit of output tend to rise towards the end of the growth phase of the cycle because of high demand for labour and imported materials relative to supply. This can be observed as a shift in the relative bargaining power between employers and their employees and measured in terms of higher real wages and lower labour productivity. Later on in recessions the opposite happens. Prime costs tend to fall, at any rate when expressed relative to the selling prices of the corporate sector's output, as employers regain their bargaining strength when unemployment rises. Profit margins will not move in line with the cycle, but will tend to lead it somewhat. A recession margin can only help to restore the profitability, and the margin of profit to turnover, of firms if wages and capital values are reduced. Lower wages mean higher profits per unit to the individual firm while lower capital values mean higher returns to the new owners of shares, while the existing shareholders take the loss in their capital assets.

Sherman (1991) and Weisskopf (1994) discuss the behaviour of aggregate business profits over time. During periods of expansion prices rise at first with costs lagging, giving firms an opportunity to increase profit margins on an increasing turnover. This provides a twofold source of increased mass of profits. Towards the end of the expansion, however, profit margins tend to be squeezed. Slower demand growth prevents firms from passing on all prime cost increases, which reach their maximum rate at this point, and turnover grows at a reduced rate. As a consequence the mass of profit tends to stabilise or even fall. In the crisis phase, at the onset of recession, the mass of profits falls sharply, mainly because of lower turnover, though also perhaps because prices fall relative to prime and capital costs. In the recession phase profit margins tend to widen again, as prime and capital costs fall relative to prices, offsetting the effect of falling sales on the mass of profit.

The profits squeeze towards the end of the expansion phase causes a fall in investment and heralds the start of recession. Likewise the early recovery of profits in the recession causes increased business confidence and can lead to a revival in investment, bringing the recession to an end. The effect of profits on investment arises both because current profits are taken as the best predictor of the likely future profitability of investment, and because current profits are used to finance investment. Firms are assumed to be either reluctant or unable to borrow without limit or without reference to the level of their *internal* sources of funds (retained profits). Sherman's key assumption is that the business sector as a whole operates cost-plus pricing but with mark-ups that vary over the stages of the cycle.

The short-run rate of growth of total income depends upon the rate of investment k, because of its effect on aggregate demand. The rate of net investment k, is the rate of accumulation of capital stock and depends on the rate of utilisation of the existing stock and the rate of profit in the previous time period. That is, the more fully utilised the capital stock (which implies less spare capacity), and the higher are recent profits, the higher will be the rate of investment. Now, the rate of utilisation of existing capital stock will increase if utilised capital stock is increasing faster than the rate of increase in the total capital stock. The rate of increase in the total capital stock is the rate of net investment. The absolute utilisation of the capital stock will depend upon the rate at which income is increasing, and on the output to utilised capital ratio.

The rate of profit depends on technology, the utilisation rate of capital, monopoly power and the relative bargaining strengths of participants in the economy. Variable direct costs per unit of output are assumed to be constant with respect to volume of sales. Prices are assumed to be set by multiplying costs by $(1+d)$, where d is an average mark-up. The size of the mark up in any industry depends upon the degree of monopoly in that industry. The degree of monopoly power is measured by the degree of seller concentration, the degree of price collusion and (inversely) the product price elasticity of demand (Cowling, 1982; Kalecki, 1971).

The average degree of monopoly is largely determined by underlying, structural features of the economy, but will also be positively dependent on the rate of utilisation of the capital stock. It is in this way that, in Cowling, the stage of the business cycle has its effect on the mark-up. For instance, as an economy begins to expand, and the rate of utilisation of the capital stock increases, rising consumer confidence reduces the price elasticity of demand enabling firms to raise their prices and hence their mark ups. We will depart from this assumption, by making the average mark up also dependent on the relative power of capitalists, reflecting the idea of profit margins on direct costs being squeezed towards the end of booms.

The influence of technology on the productivity of inputs, is fairly straightforward. The more superior the technologies becoming available, the less the direct variable input (labour and imported materials) cost to firms of a given increase in output volume, and the bigger the saving in direct cost relative to the capital cost of the technology. It raises, for example, labour productivity relative to the wage rate. Booms tend to increase productivity, following Verdoorn's Law, according to which productivity growth is a positive function of output growth, though forces exogenous to the business cycle have a greater influence.

Relative capitalist power, comprises mainly two elements. Relative capitalist power firstly depends on the *terms of trade* between an economy's firms and the rest of the world. The terms of trade is the ratio of the input cost required to produce the number of units of output needed to exchange for a given unit of imported input. It is the amount of domestically produced goods or services required in exchange for a given quantity of imports. Secondly, the relative power of capitalists depends on the class relationships between employers and their employees. This is also partly a matter of the *terms of trade* for labour power in the labour market. In this case it is the size of the real wage, measured as a quantity of goods, relative to the amount of work done, and hence goods produced, per unit of labour time. We argue that as the expansion phase of a cycle continues, the relative power of the capitalists shrinks, whilst recessions restore it.

The business cycle is driven by changes in investment. Investment depends on profits and the behaviour of profits is cyclical in nature. To summarise the argument so far, the movement of the rate of profit over the cycle is a matter of the balance between countervailing forces. During an early expansion phase of the cycle, lower price elasticities of demand and higher capital utilisation rates tend to raise the rate of profit. Productivity tends to increase as long as expansion of the market absorbs the extra output of additional capacity, thus fulfilling investment plans. This is in part because of Verdoorn's Law, which states that productivity rises with output. Productivity growth is partly endogenous to the model of the rate of change of output and investment. The rise in productivity provides an opportunity to raise the rate of profit. However, if increased capacity is greater than the market can absorb, productivity will suffer and the decline in productivity and capital utilisation will be followed by closures and redundancies. At certain phases of the cycle, such as at the peak, a loss of relative class bargaining power tends to lower the rate of profit.

Cycles

The idea of cycles in the values of any time series data is an important one. In its strong form, it implies deterministic cyclical effects of known duration. That is, the average or normal length of the cycle is known and fixed, and cycles follow one another, as it were, mechanically. This requires there to be some theorisable force at work dictating this regularity. Of course, individually cycles might depart from their predicted length, but these departures would be required to be relatively small.

On the other hand, in its weak form the idea of cycles simply implies that, when we remove the long run trend from any set of data, what we have left displays a cyclical rather than a random pattern. That is, if one year lies above the trend, it is likely that so too will the next year. Variance from the trend goes through a *cyclical* rhythm, accelerating and decelerating in a regular and smooth way. These cycles may be of highly variable length. Indeed a cycle period may be identified and measured simply by the fact that it is not possible to decompose it into shorter cycles. Alternatively, we may feel justified in ignoring minor peaks and troughs and subsuming them within a more dramatic longer cycle.

Thus (Figure 10.3, p. 244) construction output began a cyclical fall in 1973–4 until 1980–1, a fall which was interrupted for one year only in 1977–8. The data can be interpreted in different ways. For instance, one analyst might feel justified in ignoring the peak in 1978 even though it was higher than either the preceding or the following year and was, therefore, technically a turning point. Such an interpretation might be justified by adopting a long period view, and using annual data, seeing the 1978 peak as a random blip effect, and identifying one long cycle running from a peak in 1973 to a trough in 1981 and then a peak in 1990. At the same time, another analyst might stand upon strict definition and see three cycles in the same period, namely 1973 to 1978, 1978 to 1984, and 1984 to 1990.

Trend and cycle

Actual time series data for any variable will show changes from one year or period to the next. In the search to find a pattern in the data various methods of statistical analysis are used. First, there is a trend effect. One method used to find the trend is to take data for a large number of periods together, plot them on a scatter diagram and then draw the *line of best fit*. This produces the linear trend line which minimises the variance. Exceptionally the trend could be a zero growth rate, in which case the trend value would not tend to change from one year to the next. However, normally, the trend line, drawn thus, will have either a positive or a negative slope, a positive or negative trend rate of growth.

Alternatively, using he *moving average* method to measure the trend, requires that we first know the normal length of the cycle in the series. Suppose the cycle to be 15 quarters. One would then construct a 15-quarter moving average trend as follows. From the first available 15 observations (quarters) one calculates an average. This is then attributed to the 8th quarter when plotted as a graph. Then, the earliest observation is dropped and the next one added, and a new average calculated for the 2nd-to-16th quarters. This is then shown against the 9th quarter; and so on. One practical disadvantage with this method is that the trend can never be measured right up to the latest date for which there is data, for reasons which can be immediately seen. However, one advantage over the straight line of best fit method is that it allows for changes over time in the slope of the trend. In effect this moving average method *smooths* the original data. If there is no cycle of regular periodicity shown by the data, the moving average method is inappropriate.

A simple and practical procedure is to fit a straight trend line to the data and then inspect the variances of the actual observations from the trend for serial biased error. That is, if, as time passes, the trend so drawn tends increasingly to lie below the actual observations, so that the errors are mostly in one direction, we should conclude that the true slope of the trend factor has in fact increased, as compared with the earlier period. Because, as well as trend and cycle, there are random effects, there will always be some errors as measured by differences between predicted and actual values. If these errors are randomly distributed, the models used for drawing trend and cycle are acceptable.

Aside from the moving average method, there are three broad types of statistical strategy for dealing with trends with changing slope over time. The first simply identifies, by visual inspection of the diagram, the point at which the trend seems to have shifted, and divides the total period into two. A separate straight-line trend is then fitted to each sub-period. This is particularly sensible if one has some good reason to expect an underlying change in the behaviour of the variable at about the time of the break between the sub-periods. Many economists break their time-series in the early 1970s, because historical knowledge tells us that this marked the end of one economic age and the start of another. This technique is based on hindsight and has the disadvantage that quite a long time must pass before one can tell that the old trend no longer fits the data. This method is of no use for purposes of prediction.

The next simplest approach is to try to fit a non-linear but mathematically simple and smooth function to the data – for example, an exponential function. This retains the advantage of the single linear function of being projectable forward, in a deterministic way, to yield predictions of future trend values of the variable.

Finally, there are techniques for generating continuously changing trend curves. These use a statistical *filter* technique to eliminate random and cyclical effects from the data, and produce trend curves which look like (but are generated quite differently from) the n-year moving average curves, which they have replaced. These are known as structural time series models. The trend curves produced using, for example, the Kalman Filter technique (Harvey and Jaeger, 1993; Canova, 1993) make no presumption that the future slope of the curve is known, in the sense of being determined by its past slope. They smooth the raw data to an extent which depends upon the length of the data period relative to the length of the cycle period. Ball and Wood (1994A) apply these methods to long series of data for construction demand.

Bails and Peppers (1993) provide a clear and straightforward practical guide to the range of simpler methods, and equips non-mathematical readers to undertake such analyses themselves.

Business cycles

It is well established that the GDP follows a cyclical pattern of fluctuation about its trend, and so, therefore, do many other economic variables with which it is linked. These GDP cycles, known as business cycles, may be rather short, typically around five years in length from trough to trough (though their length is variable), and of highly variable amplitude. Sometimes they are very weak, but sometimes very strong. Also, sometimes the downward part of the cycle is dwarfed by the recovery with many fewer quarters of recession than of expansion, in each cycle. In other cycles the period of recession may be longer than or similar in length to the period of recovery. Clearly, if GDP is cyclical then this alone may be sufficient to explain the existence of a cycle of similar duration in construction demand. It is therefore with the business cycle that we must begin our investigation of construction cycles.

Whilst all economists agree that business cycles occur, they divide into those who view them as caused essentially by lagged responses to exogenous shocks, and those who believe the economy has its own endogenous forces at work that will cause cycles even without exogenous shocks. We take the latter position. We therefore use a simplified version of the trade cycle, which excludes the random exogenous shocks to the cycle which do indeed cause many of the fluctuations in the actual empirical data. These shocks include wars, elections, industrial unrest and government policies.

In order to keep matters fairly simple, we will also set out the model as for a closed economy, without imports or exports. Evidently, in actual, open economies booms and recessions in other, trading-partner economies can affect the business cycle in any one national economy. If different countries' cycles are synchronised, they will reinforce and exaggerate each other, whereas if unsynchronised they will dampen each other.

The 9-stage business cycle: a method of description of business cycles

Sherman (1991) describes a business cycle as a period of expansion in economic activity followed by a period of contraction. Economic activity can be measured using statistical indicators of, for instance, output, employment and profits. Reference cycles are general cycles relating to the average of all variables in the whole economy. However, it is often useful to compare the performance of a specific variable such as the output of the construction industry or changes in employment to changes taking place in the rest of the economy. A specific cycle may produce a lead or lag indicator. For instance, the specific cycle of profits leads the reference cycle, dropping ahead of the economy as a whole, while interest rates tend to be a lag cycle, only dropping after the rest of the economy has entered a decline.

According to Wesley Mitchell, there are four phases in each business cycle, called in turn, the recovery phase, the period of prosperity phase, a crisis phase and depression or, if mild, a recession phase, before a recovery begins the next cycle. Cycles are here measured from trough to trough.

For the purpose of detailed analysis and comparison of cycles, each cycle can be further broken down into 9 stages. Beginning in a trough, stage 1 is the start of the cycle. Stage 5 is the peak and the cycle ends at stage 9 at the second trough. Stages 2, 3 and 4 are periods of expansion and are defined to be equal in length and usually last longer than stages 6, 7 and 8, which are periods of decline.

These cycle stages can then be used in conjunction with cycle relatives. This method, used by the influential US National Bureau of Economic Research (NBER) works by first identifying cyclical turning points. With the start and end of a cycle thus defined, a cycle-average value is calculated, from all the quarters comprising the cycle. Each quarter's value is then measured as a *cycle relative*, or ratio of the actual value to the cycle-average value. Using cycle relatives, cycles of different economic indicators or industries can be compared. For example, cycles in interest or profit rates can be compared with cycles in GDP, not only in their overall amplitude but also in their detailed *shape* or sequence. If underlying trends are very powerful relative to cyclical movements, the cycle-average loses its value as a concept if used with un-detrended data. However, the method has the merit of being equally applicable to mild recessionary cycles and strong depressionary cycles. This is because it does not require an absolute drop in the value of a variable, for example GDP, to identify a recession. It is sufficient that the value drops below the cycle-average.

Table 10.1 illustrates the difference between cycle relatives and trends. Suppose GDP starts a 14-quarter cycle with a *trend* value of 100 and ends it with a trend value of 198 based on an average compound growth rate of 5 per cent per quarter. Now, let us imagine two scenarios. In one, the *actual* value at the beginning of the cycle is 90, and after 6 quarters (i.e. in quarter number 7) growing at approximately 9 per cent per quarter, GDP reaches a value of 151. This is well above its trend value of 134, and also above the cycle average of 149 for this scenario. Thereafter, in this scenario suppose GDP continues to rise but more slowly, below the trend rate of growth, reaching 180 by the last quarter. Although there has been a deceleration, there is no absolute fall in any quarter. Now, in another scenario, suppose growth was similar but slightly faster, at 10 per cent, up to Qtr. 7, but then as follows: a continued growth at 10 per cent to, say, 192 in Qtr. 9, before falling back to 160. In this second scenario we would have a clear *absolute* peak turning point and subsequent contraction.

The change in cycle relative from one quarter to the next is the growth rate. Because cycle relatives are measured relative to their mean value over the cycle, if there is an underlying growth trend affecting the variable, cycle relatives will tend to begin each cycle with lower values than they end it with. In absolute terms, the value at each successive trough will tend to be higher than at preceding troughs.

In reality, business cycles usually begin with relatively rapid growth and decelerate towards the peak after which decline is relatively rapid for a few quarters, thereafter slowing down until the trough is reached. The trough is, by definition, followed by recovery.

Finally, the growth and decline which takes place in each cycle can be measured, using the 9-stage method. Cycle amplitudes measure the difference between the cycle relatives at stage 1, during the first trough, and at stage 5, at the peak. The difference is a measure of expansion. The contraction amplitude is measured by subtracting the cycle relative at stage 9 from the cycle relative at stage 5.

Table 10.1 Schematic illustration of trend, differences and cycle relatives

Quarter	1	2	3	4	5	6	7	8	9	10	11	12	13	14	15
Scenario 1[2] (+9% then slow growth)	90	98	107	117	127	138	151	155	159	162	165	168	172	176	180
Cycle relative (S. 1)	.60	.66	.72	.79	.85	.93	1.01	1.04	1.07	1.09	1.11	1.13	1.15	1.18	1.21
Difference from trend (S. 1)	−10	−7	−3	+1	+5	+10	+17	+14	+11	+7	+2	−3	−8	−13	−18
Trend (+5%)	100	105	110	116	122	128	134	141	148	155	163	171	180	189	198
Difference from trend (S. 2)	−10	−6	−1	+4	+10	+17	+25	+34	+34	+25	−3	−9	−20	−29	−38
Cycle relative (S. 2)	.62	.68	.75	.82	.90	.99	1.09	1.20	1.32	1.16	1.10	1.10	1.10	1.10	1.10
Scenario 2[1] (+10% then fall)	90	99	109	120	132	145	159	175	**192**	170	160	160	160	160	160

Table 10.1 Continued

	Cycle average
Scenario 1 (+9% then slow growth)	149
Trend (+5%)	144
Scenario 2 (+10% then fall)	146

Notes:

1. In Scenario 2, the absolute peak turning point is shown in bold. Such a peak will also be the peak in the value of the cycle-relative.
2. In Scenario 1, there is no such turning point, but the point of deceleration (maximum slope of growth curve) is shown in italic. Both absolute value and the cycle relative necessarily have maxima, but these are not peaks or turning points, as they are not both preceded and followed by lower values. In S.1 presumably Qtr. 16 would show a value above 180.

This then leaves the question of whether construction output will fluctuate in time with GDP, or in a leading or lagging relationship. Note that the accelerator-multiplier model of the business cycle, as outlined above, does not really deal in time lags. In fact, production of new buildings and works takes a relatively long time from the investor's decision to spend to the delivery of a finished, useable product. As a result, investors may try to anticipate increases in consumer demand, rather than wait for them to happen. If they do not do this, they may find themselves in the position of those UK brick manufacturers whose new brick works finally came on stream, ready to start production, just as the slump in construction output started in 1990. The brick producers had waited for actually observed building demand to increase for perhaps two years from its 1985 plateau before deciding to invest in extra capacity (phase 6). It then took a further two or three years to go through all the planning, design and construction phases for the new brick works. Thus, even though five years, 1985–90, was an exceptionally long upswing in a business cycle, it was still not long enough as far as those firms were concerned.

This issue of time lags and the need to act on the basis of expectations is at the centre of a more genuinely Keynesian, or 'post-Keynesian', model of the business cycle, (Chick, 1983; Arestis, Chick and Dow, 1992). The key actors in the construction demand process acting upon expectations of future demand are the speculative developers.

Construction tender prices are affected by contractors' expectations of future direct costs, and even opportunity costs of management resources. Contractors assume that costs per unit of input will be affected by the level of demand. It is debatable, though, whether raising tender prices in anticipation of cost increases serves to choke off anticipated excess demand. When clients' other expectations are still optimistic, it may be that they are not induced to make major reductions in volume of construction demand in response to tender price increases. After all, many clients are themselves speculating, and will not be deterred so long as they expect property prices to increase faster than construction prices. Many other clients are engaged in investment decisions for whole projects of which construction will account for a relatively small part of the total project cost, which includes the cost of machinery and, more importantly, operating costs. Realistic project appraisal models reveal that expected returns are often relatively much less sensitive to changes in expected construction cost than to changes in expected interest rates or to changes in expected demand for output, as measured by sales volume and selling prices.

One key aspect of the post-Keynesian view is that, because future prices are uncertain, firms can neither bargain nor calculate in *real* terms. Every transaction becomes a speculation, in that the purchasing power of the money to be paid or received cannot be known for certain at the time. The longer the time lag between, for example, selling some output and using the proceeds of that sale to purchase production inputs, the greater the uncertainty. The time lag arises both because payment is not simultaneous with sale, and because money received will be held initially in liquid form, prior to any decision on how to spend it.

On the other hand, it is certainly convenient to theorise *as if* wages, in particular, were bargained over and set in *real terms*. Wage-earners (and unions) try to predict the real purchasing power of money before setting their *asking wage* in money terms, although they may of course get their predictions wrong. In collective bargaining systems, money wages may then be fixed until the next bargaining round, or there may be some form of indexation, guaranteeing purchasing power. In individual bargaining systems, with mobile labour, asking-wages may continually adjust to new information or expectations about prices and hence purchasing power.

Employers, on the other hand, think of the wages they are prepared to offer either in money terms, or relative to the expected price of their own product. Employers have proved much more willing to accept *ex post* or compensatory indexation, in which this year's wage increase is tied to last year's actual rate of price inflation, than *ex ante*, open ended indexation, in which wages are increased monthly, or quarterly, as the latest Retail Price Index figures become known. The former commits them to a money outlay known in advance, whereas in the latter it is unknown and uncertain. Tying wages to the previous year's profits is, in effect, a way of linking them to last year's own-product rate of price increase.

However, if all transactions are based on uncertain expectations, it follows that, if and when those expectations are falsified by events, all economic actors will then seek to adjust or alter some of their decisions. These responses to failed expectations can, if shared by enough important actors, be the engine of disequilibrium, and the cause of turning points in the trade cycle. Keynes reserved the term *speculator* for those who neither make consumption nor production decisions. Instead, they live by continually adjusting their holdings of assets, from one type of asset to another, in the light of changes in expectations, and doing so somewhat before the general run of producers and consumers change their expectations. Speculators are, as it were, professionals in the expectations business. Whereas, Keynes thought, most ordinary producers and consumers are more or less continually taken by surprise by the divergence of actual outcomes from their expectations, and have to react to this after the event. Hence, for example, one fundamental difference in the real behaviour of markets in financial assets compared, to for instance, labour markets or product markets is that financial markets, but not the others, are dominated by speculators.

However, since Keynes wrote, real estate or property has come to be treated in financial asset markets as if it were in fact a financial asset. Financial institutions hold property let on long leases as part of their portfolios, alongside cash, loans and equities, its rents being treated as equivalent to dividends on equities or interest on loans. For this reason, the property market, and hence property prices, behaves much more like a financial asset market than a product market. In the property market there is continual speculative portfolio-adjustment in the light of continuously adjusting expectations. Property demand and supply cycles result in unexpected and large changes in the rates at which the building stock is utilised or occupied, and these in turn cause major unexpected changes in the market values of those buildings as assets in the books of their owners. Speculative property markets can then have their own causal effect on real market demand for the product of the construction industry, and can be a cause or determinant of the cycle in construction demand.

These fluctuations in the property cycle have been accounted for as follows, by Bon (1989). The rate of capital utilisation declines whenever there is excess capacity. This occurs because fixed capital assets such as plant, machinery and buildings are durable and relatively inflexible in terms of their potential output. As a result the capacity of fixed assets is relatively unresponsive to changes in output and when output declines spare capacity is created in the short term, until sufficient assets are retired. The utilisation of buildings during trade cycles can be seen not only in terms of fluctuations in occupancy rates but also in terms of variation in rents.

Even when the rest of the economy had begun to recover, Lewis (1965) noted, recovery in construction was often hindered because financial institutions were reluctant to increase the amount of buildings and property available before rents and property prices had fully recovered. Broadly, Lewis' points remain valid today. Rents, and house prices, remain downwardly sticky in property market recessions. It is not that they do not fall at all, but that they do not or cannot fall to a market-clearing level, at which any available property is let relatively easily at a known going rate and rents are firm. Until the property market recovers, UK financial institutions see construction and property development as high risk ventures. Nevertheless, because of the deregulation and globalisation of financial markets, no group of lenders can any longer restrict the supply of funds to control property markets in the UK. However, the same effect can be achieved through influence on public policy. In the newly restrictive UK town planning development regime introduced in the 1990s, with respect to out-of-town retail developments, it may be possible to detect some pressure from nervous financial institutions heavily tied into existing high street retail property.

New technology may also reduce expected future space requirements. For instance, new computer technology will enable people to work from home or on the road, without the necessary requirement of a permanent personally designated office base. Other factors which influence firms' decisions about capital investment include the current and expected rate of interest and foreign penetration of domestic markets as well as opportunities for exporting future output. Finally government policies regarding foreign exchange rate, taxation and spending will all influence business confidence. As a result, building sector demand will not respond mechanistically to changes in consumer or even total investment demand.

To return to our model, it can be seen that, depending on the rates of decline and recovery in the economy and the rate of decline in the built stock, once a trough has been reached, the first phase of recovery in building demand will come in the form of increased maintenance and refurbishment. This will be followed, after a time lag, by an increase in demand for new building work as a point is approached when the existing stock is about to be fully utilised. The exceptions to this rule are when the existing stock is not in a desired location or the building types available do not conform and cannot be converted to the requirements of any new uses expanding firms may need – i.e. where there has been a fundamental qualitative change in the basis of demand.

Measurement of construction demand

There are several possible ways of trying to measure aggregate construction demand. In microeconomics *demand* refers to a set of possible price-quantity combinations. In written usage, the term 'demand' tends to be used as shorthand for the *quantity* that will be bought at a certain price, as for example in the term *price elasticity of demand*. In macroeconomics *demand* refers to a value of expenditure, found by multiplying price by quantity. If prices are assumed to be stable, then demand in this sense comes to refer to the quantity of output that is actually bought. It is in this way, for instance, that employment is said to be a function of aggregate demand, since employment is in fact a function of the quantity of output produced, and not directly a function of the value of expenditure or value of output.

Now, the nearest thing we have to a statistical measure of *expenditure* on construction is the *GDFCF series*, which does indeed measure the value of expenditure on fixed capital formation by purchasers of new capital goods. However, by definition it does not cover or include expenditure on repair and maintenance of the built stock, and thus completely misses one significant component of aggregate construction demand.

For estimates including repair and maintenance demand, in the absence of timely expenditure statistics, we are forced to use *construction output value* as our second possible measure for repair and maintenance projects; given R & M projects' short duration, this is perhaps acceptable, since almost all output will in fact occur in the same short time period as that in which the decision to undertake expenditure was made by the customer. However, for new construction projects the time lags between demand decisions and output may be considerable.

However, as we have pointed out elsewhere, it is a peculiarity of new construction demand that purchasers have to largely commit themselves to the decision to purchase a certain quantity of output *before* they know what the price will be. They may well have access to a price prediction, of course. But a tender price can only be found after:

- a design has been commissioned (at considerable expense to the purchaser), and
- an invitation to submit tenders has been issued.

An invitation to tender typically implies a commitment to accept the lowest tender, regardless of its absolute value. If a regular client frequently breaks this implied commitment they will find it difficult to persuade firms to incur costs of tender preparation and to tender at competitive prices. Nevertheless, if would-be clients are unpleasantly surprised by actual tender prices they may cancel or postpone. Only to the extent that this is so is *output* volume a reasonable proxy for the volume that clients *wish* to buy at the actually prevailing level of construction prices, and thus a proxy for the concept of demand volume.

The third set of statistics sometimes used to measure construction demand is the *orders series*. These measure the value of new contracts placed with construction firms by clients in a period. Deflated for price changes, the orders series can be used as a proxy for change in the volume or quantity of construction that customers wish to buy at the prevailing level of prices. It has the advantage over output of not lagging so much behind the actual decision to spend, but the disadvantage that orders data are only available for non-repair and maintenance construction. The main advantage of orders over GDFCF data is that the former are published more quickly and are disaggregated by location and type of construction product ordered rather than as is the case with GDFCF by sector or industry of origin of the demand. They are thus much used by construction firms trying to observe demand changes for their particular product market.

The approaches to measuring demand given here are based on ONS expenditure data in the national accounts, DETR construction orders or DETR contractors' output. Each method provides answers to different questions regarding demand for construction.

In principle the most all encompassing measure of construction demand is the total value in current prices of actual expenditure on purchasing the products of the construction industry including all new buildings and works as well as all work to existing structures. Because this is an expenditure based measure it includes professional fee charges. It is the sum of GDFCF expenditure on net acquisition of new buildings and works, *and* consumption expenditure (by all sectors) on repair and maintenance. This is a composite measure, requiring figures for personal, government and corporate sectors' consumption expenditure on construction repair and maintenance to be identified, and added to GDFCF figures from the Blue Book. Repair and maintenance figures can be found in the output data of *Housing and Construction Statistics* published by the DETR, and for the household sector only, R & M expenditure can be found in the *Blue Book*.

Any of the approximate measures of aggregate construction demand can be converted to constant prices using the implied GDP deflator. The implied GDP deflator is the inverse of the index of average price changes for the whole economy. Measuring construction demand this way, though not a *volume* measure in the direct sense, has the advantage of showing the amount by which construction demand and output

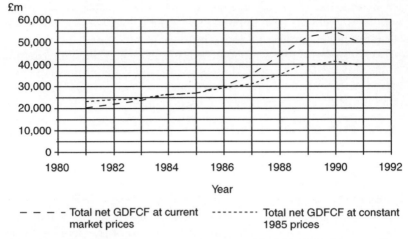

Figure 10.1   Total net construction component of GDFCF, 1981–91, expenditure based measures of construction demand
Sources: *United Kingdom National Accounts, 1992*, Tables 1.7 and 13.4.

have increased in volume terms relative to other sectors of the economy. Alternatively to the implied GDP deflator, construction price indices can be used to deflate construction demand into constant prices and thus produce a quasi-volume measure.

A narrower definition of demand, shown in Figure 10.1, looks at the value in current prices of GDFCF on new buildings and works only, ignoring the consumption expenditure on repair and maintenance. This measure has the practical advantage of being immediately available for each year in the *Blue Book*.

Because this measure can be found in the national accounts, there are corresponding ONS implied deflators for construction and it can therefore be converted to constant prices. This is also illustrated in Figure 10.1. These deflators relate to prices of dwellings and all other buildings and works. They have the advantage that they can disaggregate constant price demand for dwellings and other buildings and works whereas the implied deflator based on the DETR construction output price index for all new construction can only be used for construction output as a whole. This can be used as a comparison to the aggregate of the ONS deflators, because in principle both ONS and DETR implied deflators are attempting to measure more or less the same thing. If they diverge, one set of data must be chosen as more accurate than the other.

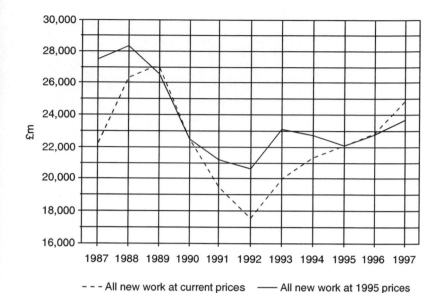

Figure 10.2 Contractors' orders at market prices, 1987–97, orders based measures of construction demand
Source: *Housing and Construction Statistics, 1987–97*, Table 1.1 (London: HMSO).

The following measures of demand are based on construction orders rather than expenditure or output. The first is the gross value in current prices of orders for new construction won by construction industry firms and is shown in Figure 10.2. (Orders data are not available for repair and maintenance work.) In principle this should equal the value of the net construction element of GDFCF, that is, after deduction of professional fees paid directly by clients. This is because there is in theory no time lag between *ex ante* expenditure and order. They are the two sides of the same transaction, so long as customers report the whole value of their investment expenditure on a construction project in the year in which the contract is signed, and report it at contract rather than final account price. In practice, as we have seen, they do neither, so GDFCF corresponds in time to output, and even lags it slightly. In principle, though, GDFCF should mirror orders albeit with a lag, because GDFCF measures the purchase and orders measure the sales side of the same transaction or contract.

In practice GDFCF and *lagged* orders are still not equal, partly because of incompleteness in the censuses of construction firms and of projects from which the orders data are drawn, and partly because some investment expenditure on new structures is not directed as demand to the construction industry at all. Some part of this demand for structures is met by industries other than construction (for instance, process plant engineering). Moreover, of the part of GDFCF (structures) that is demand for the kind of final products produced by the construction industry, some of this work is undertaken by the construction departments of the same organisations that generate the demand, or it is carried out and not recorded in the informal, black economy, or it is done by individuals as self-build. Thus, for example in 1991, according to the *Blue Book*, the structures element in GDFCF was £31.8 billion at market prices, while the gross value in current prices of orders for new construction won by construction firms was only £19.5 billion.

The incompleteness of the orders data has been demonstrated for earlier years, the 1970s, by Ball (1988). Looked at over a period of years to exclude cyclical effects, he showed that orders data systematically fell short of reported construction industry gross output data, even when both were expressed in constant prices. Figure 10.4 below shows that this is still the case, and DETR orders and output data are still not comparable.

Figure 10.3 shows the annual value of orders won by contractors at constant 1990 prices and total output including repair and maintenance of all firms in construction including DLOs.

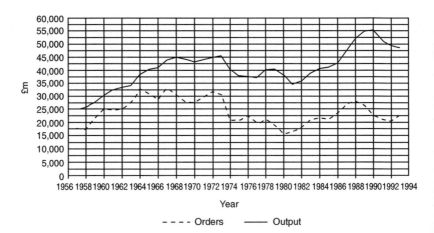

Figure 10.3   Orders and output series compared, 1957–93, at 1990 constant prices
Source: *Housing and Construction Statistics, 1994*, Tables 1.1 and 1.6.

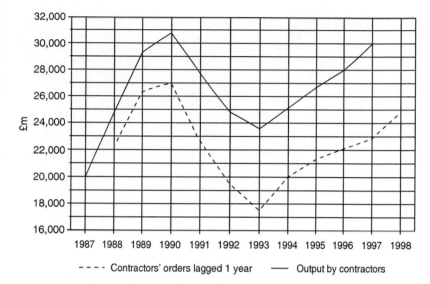

Figure 10.4 Orders lagged by one year and contractors' output, 1987–97, all new work at current prices
Source: *Housing and Construction Statistics, 1987–97*, Tables 1.1 and 1.6.

In Figure 10.4 the value of contractors' orders, and the value of new work by contractors are plotted. This shows the relationship between orders and output of private contractors. Orders are lagged by one year to take the period between orders and production into account. The figures shown in, say, 1989 represent output in 1989 and orders won in 1988. The lagged figures provide a close match, both rising and falling at similar rates and peaking in 1990.

Fixed price contracts include terms under which the contractor includes an estimate of inflation in the initial contract price, and bears this risk. When most contracts, measured according to value, are set in *fixed price* terms, then in principle, after a time lag, a certain money value of orders ought to turn into the same money value of output or *work done*. When most contracts contain 'fluctuation of price' clauses, so that the customer bears the risk of inflation, this will not be so. With normally positive rates of inflation, values of projects measured at completion or as paid for will then exceed initial contract values. Such fluctuating price contracts were common in the 1970s, but are no longer widely used.

A further reason for differences between orders and output figures lies in the strongly upward effect on out-turn prices, compared to

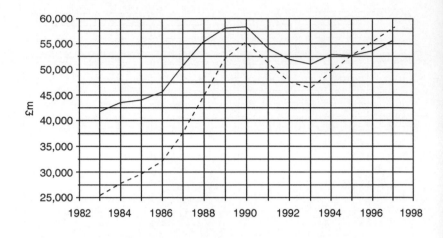

- - - - Output at current prices   —— Output at 1995 prices

Figure 10.5 Construction output, 1983–1997, output-based measures of construction demand
Source: Housing and Construction Statistics, 1987–97, Table 1.7.

tender prices, of *variation orders* arising from variations in the technical content of work to be done, announced after the date of tender award. These raise the value of contractors' output without affecting the value of orders, measured at tender price. The increase in final building costs due to variation orders cannot be dealt with even in principle by an output price index. Nevertheless, output series can be deflated using output price indices to express output in constant prices. This has the effect of eliminating 'variation of price' clauses, and is therefore in principle comparable with series for orders. See 'Appendix: Notes and Definitions' to *Housing and Construction Statistics* for an account of how this is done.

An alternative measure of demand, using constant prices, is the gross value of orders for new construction won by construction firms, deflated using tender price indices. This is also illustrated in Figure 10.2 but this measure of demand possibly suffers from the non-representativeness of the DETR tender price indices, which relate only to public sector contracts.

Figure 10.5 shows two measures of construction demand based on DETR data on contractors' output. Firstly, construction demand may be seen as equivalent to the gross value at current prices of output, known as *work done*, in the period by main contractors or house-builders. This should be the same as the value of work done by *all* firms, *net* of the value of work done for them by construction industry subcontractors. The former method is used in the project-based and the latter in the firm-based estimates of output. Where these estimates differ, in practice the DETR relies upon the project-based enquiry to estimate industry output. Value of output is here defined as the amount chargeable to customers for work done in the relevant period. It is not clarified whether this includes or excludes contractors' own estimates of the value of work done in a period but for which no valuation certificate has yet been issued on behalf of the client.

The gross value of contractors' output at constant prices shown in Figure 10.5 is found by deflating the current price of new build work using output price indices and deflating repair and maintenance at current prices using cost indices.

GDP and aggregate construction demand

Figure 10.6 shows that total construction output rose steadily until around 1968. Although 1973 is usually seen as the peak year for construction output. The rate of growth appears to have slowed down well before 1973. Certainly, after 1973, construction output began to fluctuate, falling erratically until the beginning of the 1980s, after when it rose until 1990. More understanding of the trend can be found by analysing the data in terms of new build and repair and maintenance. In the period between 1955 and 1968, the rise in construction output was mainly driven by new build, with repair and maintenance only growing moderately. After 1968 the picture changes to a period of fluctuating output, when both new build and repair and maintenance can be seen to fluctuate more than in the earlier period, with the decline in construction output being caused by the decline in new build work. Since 1969 growth in total construction output has been due to the rise in repair and maintenance work rather than new build. Indeed, Figure 10.6 shows clearly that from 1969 until the 1990s, new build construction actually fell slightly in real terms while repair and maintenance continued to rise in spite of periods of recession. These graphs support our view that following a period of growth, ending in the late 1960s, construction industry output of new build projects has fluctuated without showing a pattern of longer term growth.

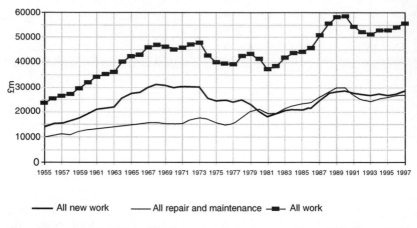

Figure 10.6   New build and repair and maintenance output, 1955–97, at 1995 prices
Source: Contractors' output, *Housing and Construction Statistics, 1999,* Table 1.6.

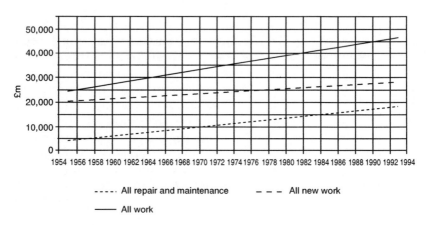

Figure 10.7   Linear trends of contractors' output, 1955–93
Source: calcalated from data in *Housing and Construction Statistics.*

Analysis of the linear trends in Figure 10.7 over the period confirms that total construction growth from 1955 to 1993 is similar to the growth of repair and maintenance.

In fact the linear trend of new build construction is even clearer when split between the periods 1955–68 and 1969–93. In Figure 10.8,

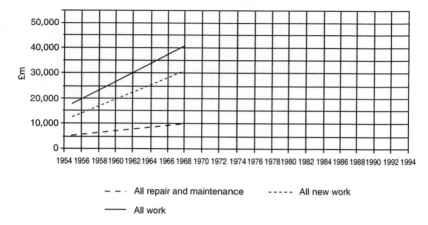

Figure 10.8 New build and repair and maintenance output, 1955–68, at constant 1990 prices
Source: calculated from data in *Housing and Construction Statistics.*

we see that growth in construction new build due to housing programmes and other urban developments was sustained up to 1968. The rate of growth of construction was so rapid that if the trend had continued at the same rate, total construction output would have reached £80 billion by 1992, instead of the £45 billion actually achieved. In fact, in Figure 10.9, the linear trend for the period since 1969 shows a much lower rate of growth in overall construction output and even a slight decline in new build output. It does not, of course, follow that new build construction will continue to decline *ad infinitum.*

Using all the above techniques applied to the construction industry and GDP, the performance of the construction industry can be seen relative to the rest of the economy and also over time. If the low-growth fluctuating description of contemporary construction demand is correct, then we would expect to find most of the following:

• comparing cycles between 1950 and 1973 with those since 1973, contraction and expansion amplitudes in the construction sector would both have grown. Moreover, the size of contractions would have grown relative to the amplitude of expansions. The result of these changes would therefore be greater fluctuations together with a low growth trend, or historical 'stagnation'.
• comparing construction with GDP, construction cycles would have the greater amplitude.

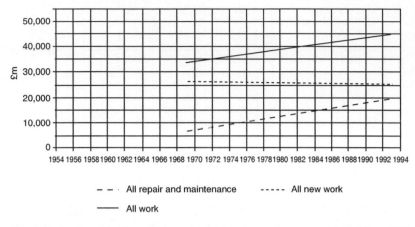

Figure 10.9   New build and repair and maintenance output, 1969–93 at 1999 prices.
Source: calculated from data in *Housing and Construction Statistics.*

Figure 10.10 shows the relative size and growth patterns of construction and GDP. From 1971 up to 1993, relatively little change in construction output was sufficient to support growth in the rest of the economy.

However, Figure 10.11 shows that when construction output is magnified 10 times to make it visually comparable with GDP, it can be seen to vary far more than the economy as a whole. This is not altogether surprising. GDP being the sum of its component industrial parts, the whole is always likely to fluctuate less than any one of its parts, because those parts are not likely to be 100 per cent synchronised in their fluctuations. A kind of *smoothing out* occurs in their summation.

Nevertheless, during the years 1973–75, 1979–81 and again from 1989–93, when the economy was in recession, decline in construction was more rapid than for the economy in general. The oil crisis, uncertainty, lack of confidence and high interest rates appear to have depressed construction between 1975 and 1977 although the economy itself managed to grow slowly. However, during the periods of economic expansion from 1977 to 1979 and 1981 to 1989, especially from 1986 to 1989, construction output grew far more rapidly than did GDP. These expansions in construction were more rapid than the rest of the economy as a whole and for this reason the increase in construction output had to be absorbed, perhaps with difficulty, by the rest of the economy.

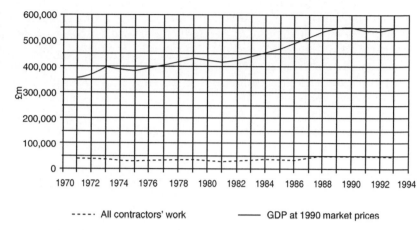

Figure 10.10 Construction output and GDP, 1971–93, at constant 1990 prices
Source: calculated from data in *Housing and Construction Statistics*.

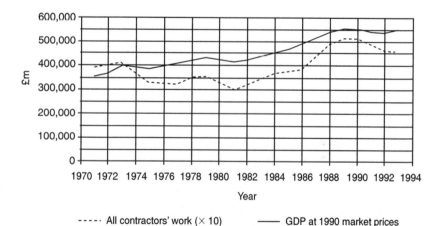

Figure 10.11 Transformed construction output and GDP, 1971–93, at constant 1990 prices
Source: calculated from data in *Housing and Construction Statistics* and the *Blue Book*.

Another issue is not so much the amplitude of variation in construction output compared to the GDP but the amplitude of variation of construction output compared to the output of other similarly-sized industry components of GDP.

We expect to find a lower trend rate of growth and greater fluctuation in GDP in the post-1970 period than earlier, and hence a lower growth and more instability in the derived demand for construction. We also find, as we would expect, a lower trend rate of growth for construction than GDP. Figure 10.12 illustrates that between 1971 and 1993 the long term growth rate of construction was lower than the economy as a whole.

The slower growth rate in construction over such a long period of time has resulted in a tendency for the share of construction in GDP to fall, as shown in Figure 10.13.

Product cycles and the trend rate of change in aggregate construction demand

To understand what causes demand for goods and service to vary over the long run, one recent approach is to consider a model in which changes in fashion and technology lead to markets for new products emerging and growing while older products become obsolete and their markets stagnate and then decline. Considered over time then, each product will show a profile of initial fast growth in output, followed by a slowing down, then stagnation and eventual decline.

It might be thought that construction, since in one sense it caters to an eternal human need, for shelter, is immune from this process. However, it can be shown that this is not the case. Demand is not for abstract shelter but for specific categories of building or works, which we shall call building types, each based on a combination of function, form, technology and property type. Thus, transportation, in the abstract, is a need as old as large scale human society. However, that does not mean that new demand for a particular transport built type, such as railways or motorways, does not follow a product cycle.

In aggregate, construction demand will tend to fall over the very long run if there are not enough major new constructed product types emerging to replace those for which demand is stagnating or falling.

Government statisticians capture this process by from time to time recognising, and beginning the separate measurement of, the production of new building types. One recent example was the production of off-shore structures (for oil and gas exploration and extraction). However sometimes the new product type is simply concealed within a more aggregated category containing several building types. For example, purpose-built housing for the elderly could be considered a

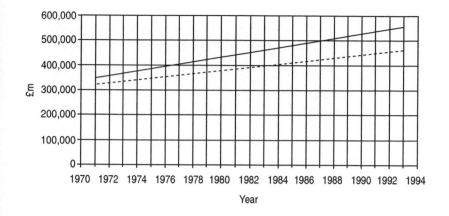

Figure 10.12 Linear trend of construction output and GDP, 1971–93, at constant 1990 prices
Source: calculated from data in *Housing and Construction Statistics* and the *Blue Book*.

new product type, but it is not yet recognised as such in the statistics for private housing output.

For long period analysis the official statistics are the only possible source, and so we will have to use them. The *Housing and*

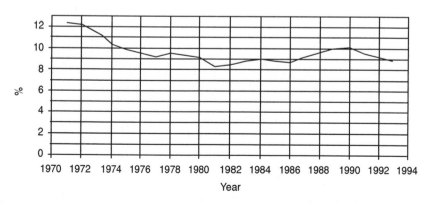

Figure 10.13 Construction as a percentage of GDP
Source: calculated from data in *Housing and Construction Statistics* and the *Blue Book*.

Construction Statistics data breaks construction output down into five types of new work and three types of repair and maintenance work:

> *New work*
>   Public housing
>   Private housing
>   Public other
>   Private industrial
>   Private commercial
> *Repair and maintenance work*
>   Housing
>   Public other
>   Private other

In 1980, *infrastructure* was introduced as a new statistical category, and in 1984 public sector housing repair and maintenance was sep-

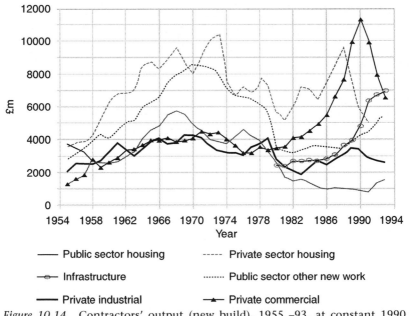

Figure 10.14   Contractors' output (new build), 1955 –93, at constant 1990 prices
Source: *Housing and Construction Statistics, 1994,* Table 1.7.

arated from private sector housing repair and maintenance. In practice data availability restricts the time series to starting in 1955.

Each of the categories shown in Figures 10.14 and 10.15 is of course, in our terms, a collection of several, perhaps many, product types. However, if we compare data for the post-war 'golden age' for construction demand with the last twenty years or so, we find that almost all of these eight categories of work showed clear trend towards growth in output in the period from 1955 to 1973, whereas in the 1974–92 period many had moved into showing a declining or stagnant trend.

The variety of patterns and the extreme variation within given time series of the different categories of work demonstrate the diverse nature of the construction market. Each time series can be analysed in its own terms. Different causes and effects influence each type of construction demand. It is not possible to predict accurately the behaviour of one type by inferring from another.

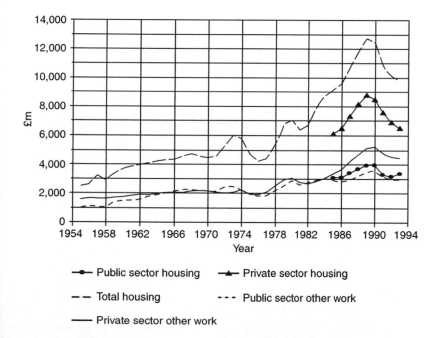

Figure 10.15 Contractors' output (repair and maintenance), 1955–93
Source: *Housing and Construction Statistics, 1994*, Table 1.7.

Economic development and the trend in construction demand

Rostow (1971) describes separate stages in economic growth. The first is the traditional stage, which is characterised by an unchanging rural economy, based on subsistence agriculture. This is followed by the establishment of preconditions for take-off through the dismantling of the traditional system. A take-off in agricultural output leads to a rise in living standards, which in turn leads to increased demand for industrial products. The economy then grows until the rate of growth slows down as the economy approaches maturity. The last stage of economic growth is characterised by high mass consumption.

The stages of economic growth can also be distinguished in terms of economic structure. Development can be characterised as a shift of resources, mainly in the form of people moving from farming the land to manufacturing in the towns and cities, the transformation of a rural agricultural economy into an urban industrial economy and then into an urban service economy. Development is a process of structural change.

Many economic historians do think in terms of a succession of determinate stages of economic growth, of which the latest phase is described as 'maturity'. Implicit behind the use of this concept is the idea that all fundamental shifts leading to the full development of a completely *modern* economy have now occurred, in countries such as the UK, with the implication of less dramatic economic change, and probably lower growth, in the future. However, it is always tempting, for those fundamentally content with the status quo, to believe that one's own generation is living, uniquely, 'at the end of history'. It almost certainly is not the case. It is just that we cannot at the moment clearly foresee what dramatic future shifts there will be.

Related to this notion of maturity is the idea that, as the economy shifts from being *developing* to being *mature* its relative need for investment in additional built stock, compared to other things, will decline, thus reducing construction's share of GDP. Indeed such a trend in the UK was noted earlier in this chapter.

There is a body of literature reviewed most recently in Bon and Crosthwaite (2000) which purports to detect such an inevitable eventual decline in the share of construction in GDP, as economies get wealthier and more mature. It is indeed observable that countries undergoing successful *development* show shares of total investment in

GDP as high as 40 per cent, whilst the 'most developed and mature economies' of today, such as the US and much of Western Europe, show ratios less than half that. Thus, even if construction's share of total investment were a constant, its share in GDP would still fall by half as economies pass from 'development' to maturity. One problem with this reasoning is that there is no evidence to support the idea that the investment share of GDP in the US or UK or similar economies when they were first developing (in the nineteenth century) was ever anything like as high as the 30–40 per cent of GDP shown recently by countries such as Japan and South Korea. Moreover, in the 1950s to 1970s, the countries with the highest shares of investment in GDP included Poland, the USSR and much of the rest of eastern Europe (Drewer, 1980).

In this chapter our argument has been that construction demand, especially but not only in the UK, has passed from a long period, roughly, from the end of World War II until around 1970, when the trend was steadily and markedly upward. Since the early 1970s the long run trend has been one of negligible growth in construction. Some major components of construction demand have even experienced decline in the period. These trends have been shown in the graphs.

Building cycles

In addition to building cycles arising as part of the short business cycle, and with its same short duration, many economists have also asserted the existence of somewhat longer cycles in construction demand. To avoid confusion, we propose to call the kind of short cycle we have been discussing up to now *construction business cycles*, and reserve the term *building cycles* for these hypothesised longer cycles. Building cycles are often in effect analysed as specifically housebuilding cycles, and must be thought of as superimposed on, and not an alternative to, shorter construction business cycles.

One postulated kind of building cycle appears to have a duration of around eighteen years. This is usually referred to as the Kuznets cycle. Kuznets himself only found evidence of such cycles in the period before 1940. He could not find such a cycle in the post-1940 US data. Nevertheless, to Lewis (1965), the problem is to consider why the building cycle should be so much longer than the business cycles of other sectors in the economy.

Ball and Wood (1994) advance a theory of slow adjustment of *housing* output to fundamental technological changes and changes affecting the relative real cost of housing and other commodities. There is a price substitution effect during housing booms. As house prices rise, relative to other products and services, this eventually reduces demand and ends the boom in housing as consumers switch their spending away from dwellings. As it becomes harder to sell new houses, housebuilders reduce their output or withdraw from the market. A long period of low building creates a backlog of demand, with house prices in general showing signs of hardening. This eventually attracts builders back into the market and after a time lag for production supply begins to catch up on demand once more. Such a model of the housing market ignores the role of government. In fact the influence of government on the housing market has been profound as government policies periodically shifted to and from generous subsidy, stimulating or damping demand and prices.

Ball and Wood (1994) found housing cycles of varying duration from 13 years in Germany up to 25 years in Sweden. In the UK from 1856-1992, they identified a 39 year cycle in the annual log constant price dwellings component of GDFCF. However, they found that since 1950, there has been a much shorter housing cycle in the UK of 17 years. Between the 1850s and the 1990s, according to Ball and Wood, housing investment grew annually at a trend rate of 2.4 per cent. However, they found that the annual average growth rate in the dwellings component of GDFCF declined from 8.8% in 1920–38, to 4.1% in 1947–73, and then to –1.3% in 1973–92. These longer term changes in demand can be thought of as forces causing a change in the trend rate of growth of aggregate construction demand.

If we assume periods of faster and slower trend growth tend to succeed one another in a regular way, then changes in construction demand and output can be seen as long cycles. However, there is a vigorous and interesting debate about whether growth rate long cycles, of any degree of regular length, exist, either for GDP or for aggregate construction demand (Mandel, 1980; Van Duijn, 1983; Tylecote, 1992; Maddison, 1991). Nevertheless, empirically there is no doubt whatsoever that trend rates of growth in GDP and in construction demand do vary over time.

Price fluctuations

So far, we have discussed fluctuations mainly in terms of fluctuations in the volume of demand and output. We have reviewed the data concentrating on the constant price value of orders and output series.

However, price fluctuations are also important. Changes in construction prices profoundly affect construction firms' profits, and also affect earnings per worker in the industry. Price cuts are passed back to subcontractors, eventually resulting in lower wage rates paid to building labour. Moreover, changes in property prices and land prices affect the wealth and credit worthiness of all property owners, and the profitability of all property developers.

Comparing the indices of construction output prices in Figure 10.16, their relative positions altered between 1982 and 1992. In other words they rose and fell at different rates depending on the particular circumstances existing in their respective markets. In contrast, if we now turn to the price indices of different materials a completely different pattern emerges, as shown in Figure 10.17. In this case, the price indices of a variety of materials follow a relatively similar pattern until 1990. Thereafter they exhibit a range of responses to the reces-

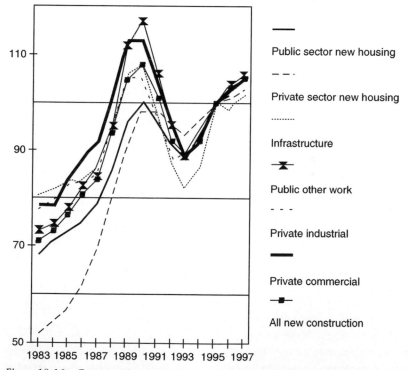

Figure 10.16   Construction output price indices, 1983–97, (1995 = 100)
Sources: *Housing and Construction Statistics*, 1983–93 and 1987–97, Table A.

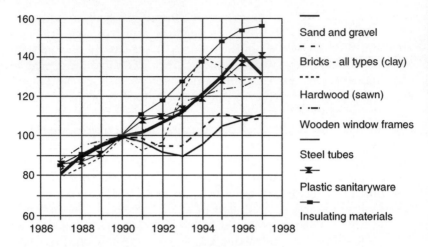

Figure 10.17   Price indices of construction materials 1987–97 (1990 = 100)
Source: *Housing and Construction Statistics, 1987–97*, Table 4.1.

sion in the industry, some continuing on a rising trend while others fell, bricks being the most volatile in the period after 1990.

By further contrast, the tender price index of all public sector contracts (apart from housebuilding) shown in Figure 10.18, peaked in 1989, the same year infrastructure output prices peaked, even though the majority of output price indices were still rising and before material prices had begun to diverge. By 1997 the index had recovered its 1989 value.

Concluding remarks

The amplitude of variations in construction output increased from 1969 to 1993 compared to the period between 1955 and 1968. However, although construction output as a whole has become more variable, the trend in construction output growth has become flat. We characterise this pattern as *stagnation-fluctuation* or *stagfluctuation*. This does not, however, imply, as seen in the data, that different types of outputs and inputs behaved in identical fashion.

By studying the data over the long run, it becomes apparent that while in recent decades the rest of the economy has grown the construction sector has not expanded in line. In the long run the output of the construction sector has been sufficient to maintain the existing

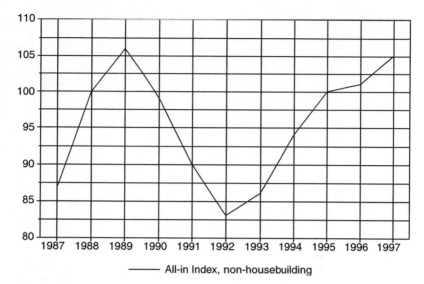

Figure 10.18 Smoothed tender price index for all public sector non-house-building contracts, 1987–97
Source: *Housing and Construction Statistics, 1987–97*, Table A.

stock, replace demolished stock and still furnish the rest of the economy with net or additional units of the built environment. These net additions to stock have it seems allowed the rest of the economy to expand without it being necessary to increase the size of the construction sector. Consequently, the construction sector has formed a declining proportion of economic activity, while at the same time maintaining its own level of output over the long run. It may be a volatile industry in the short run, but while other industries may expand and contract or even disappear altogether, the construction industry remains an essential and permanent component of any growing economy.

Bibliography

Abramovitz, M. (1968). 'The passing of the Kuznets cycle', *Economica*, pp. 45–58.
Akintoye, A. and Skitmore, M. (1991). 'Profitability of UK construction contractors', *Construction Management and Economics*, 9, pp. 311–25.
Alchian, A. and Demsetz, H. (1972). 'Production, information costs, and economic organization', *American Economic Review*, 62, pp. 777–95.
Anderson and Gatignon (1986). 'Modes of foreign entry: a transaction cost analysis and propositions', *Journal of International Business Studies*, Fall.
Andrews, P.W.S. (1964). *On Competition in Economic Theory*, London: Macmillan.
Andrews, P.W.S. and Brunner, E. (1975). *Studies in Pricing*, London: Macmillan.
Arestis, P. and Kitromilides, Y. (eds) (1989). *Theory and Policy in Political Economy: Essays in Pricing, Distribution and Growth*, Aldershot: Edward Elgar.
Arestis, P. (1992). *The Post-Keynesian Approach to Economics*, Aldershot: Edward Elgar.
Arestis, P. and Chick, V. (eds) (1992). *Recent Developments in post-Keynesian Economics*, Aldershot: Edward Elgar.
Arestis, P., Chick, V. and Dow, S.C. (eds) (1992). *On Money, Method and Keynes*, London: Macmillan.
Armstrong, P., Glynn, A. and Harrison, J. (1991). *Capitalism since 1945*, Oxford: Blackwell.
Bails, D.G. and Peppers, L.C. (1993). *Business Fluctuations: Forecasting Techniques and Applications*, 2nd edn, Englewood Cliffs: Prentice-Hall.
Ball, M. (1983). *Housing Policy and Economic Power: The Political Economy of Owner-occupation*, London: Methuen.
Ball, M. (1985). 'Coming to terms with owner occupation', *Capital and Class*, 24, pp. 15–44.
Ball, M. (1988). *Rebuilding Construction*, London: Routledge.
Ball, M. (1994). 'The 1980s property boom', *Environment and Planning A*, 26, pp. 671–95.
Ball, M. and Wood, A. (1994A). *Trend Growth in post-1850 British Economic History: the Kalman filter and historical judgement*, South Bank University, Working Paper.
Ball, M. and Wood, A. (1994B). *Housing Investment: Long-run International Trends and Volatility*, Birkbeck College Discussion Papers in Economics, 11/94.
Ball, M. and Wood, A. and Morrison, T. (1994). 'Structures, Investment and Economic growth: a long-run international comparison', South Bank University, *Working Paper*.
Ball, M. *et al.* (1998). *Economics of Commercial Property Markets*, London: Routledge.

Banwell (1964) *Report of the Committee on the Placing of Management Contracts for Building and Civil Engineering* (the Barnwell Report) (London: HMSO).

Barlow, J. *et al.* (1997). *Towards Positive Partnering*, Bristol: Policy Press.

Barrett, S.M. *et al.* (1990). 'The land market and the development process: a review of research and policy', *Occasional Paper*, 2, School for Advanced Urban Studies, University of Bristol.

Baumol, W.J. (1982). 'Contestable markets: an uprising in the theory of industrial structure', *American Economic Review*, 72, pp. 1–15.

Beer, S. (1972). *The Brain of the Firm*, London: Allen Lane.

Bhaduri (1974). *The Building Timetable: the Significance of Duration*, London: UCL.

Bishop, D. (1975). 'Productivity in the construction industry', in Turin, D. (ed.), *Aspects of Construction Economics*, London: Godwin.

Bon, R. (1989). *Building as an Economic Process*, London: Routledge.

Bon, R. and Crosthwaite, D. (2000). *The Future of International Construction*, London: Telford.

Bootle, R. (1986). *The Death of Inflation*, London: Nicholas Brealey.

Bowles, S. and Edwards, R. (1993). *Understanding Capitalism*, New York: Harper-Collins.

Bowles, S., Gordon, D. and Weisskopf, T. (1989). 'Business ascendancy and economic impasse', *Journal of Economic Perspectives*, 3, pp. 107–34.

Bowley, M. (1966). *The British Building Industry: Four Studies in Response and Resistance to Change*, Cambridge: Cambridge University Press.

Braverman, H. (1974). *Labor and Monopoly Capital*, New York: Monthly Review Press.

Brealey, R.A. and Myers, S.C. (1996). *Principles of Corporate Finance*, 5th edn, New York: McGraw-Hill.

Brenner, M. (1998). 'The economics of global turbulence', *New Left Review*, 229, pp. 1–265.

Building and Civil Engineering EDCs (1976). *Construction into the Early 1980's*, London: National Economic Development Office and HMSO.

Building and Civil Engineering EDCs (1978). *How Flexible is Construction? A Study of the Resources and Participants in the Construction Process*, London: National Economic Development Office and HMSO.

Buzzell, R.D. and Gale, B.T. (1987). *The PIMS Principles: Linking Strategy to Performance*, New York: Free Press.

Campagnac, E., Lin, Y.J. and Winch, G.M. (2000). 'Economic performance and national business systems – France and the UK in the international construction sector', in Quack, S., Morgan, G. and Whitley, R. (eds), *National Capitalisms: Global Competition and Economic Performance*, Dordrecht: Benjamin.

Canova, F. (1993). 'Detrending and business cycle facts', *CEPR Discussion Paper*, London.

Central Statistical Office (1993). *UK National Accounts, Sources and Methods*, 4th edn, London: HMSO.

Central Statistical Office (1993). *Report on the Census of Production 1991, PA 500: Construction Industry*, London: HMSO.

Chamberlin, E.H. (1933). *Theory of Monopolistic Competition*, Cambridge, Mass., Harvard University Press.

Chick, V. (1983). *Macroeconomics after Keynes*, Cambridge, Mass.: MIT Press.

Coase, R. (1937). 'The nature of the firm', *Economica*, 4, pp. 386–405.

Coase, R. (1960). 'The problem of social cost', *Journal of Law and Economics*, 3, pp. 1–44.

Cole, K., Camera T. and Edwards, C. (1991). *Why Economists Disagree*, London: Longman.

Construction Industry Group (1981). *Construction Industry into the 90s*, London: Institute of Marketing.

Coombs, R., Saviotti, P. and Walsh, V. (1987). *Economics and Technical Change*, London: Macmillan.

Cowling, K. (1982). *Monopoly Capitalism*, London: Macmillan.

Cullen, A. (1983). 'The changing interface between the construction industry and the building material industry in Britain', *Production of the Built Environment*, Proceeding of the Bartlett International Summer School, London: UCL, 4, pp. 3.45–3.53.

Cyert, R.M. and March, J.G. (1970). *A Behavioural Theory of the Firm*, Englewood Cliffs: Prentice-Hall.

Dahlman, C.J. (1979). 'The problem of externality', *Journal of Law and Economics*, 22, pp. 141–62.

Darwin, C. (1929 edn). *The Origin of Species by Means of Natural Selection*, London, Watts.

Department of Employment (Quarterly to 1995). *Employment Gazette*, London: HMSO.

Department of Environment, Transport and the Regions (formerly Department of the Environment) (annual) *Housing and Construction Statistics*, London: HMSO.

Department of Environment, Transport and the Regions (formerly Department of the Environment) (monthly) *Monthly Statistics of Building Materials and Components*, London: DETR.

Devine, P. *et al.* (1985). *Introduction to Industrial Economics* (4th edn), London: Unwin Hyman.

Dietrich, M. (1994). *Transaction Cost Economics and Beyond: Towards a New Economics of the Firm*, London: Routledge.

Drewer, S. (1980). 'Construction and development: a new perspective', *Habitat International*, 5.

Druker J. and White G. (1995). 'Misunderstood and Undervalued? Personnel Management in Construction', *Human Resource Management Journal*, 3, 3, pp. 77–9.

Drucker, P.F. (1994). *Managing for Results: Economic Tasks and Risk-taking Decisions*, Oxford: Butterworth-Heinemann.

Drucker, P.F. (1995). *Managing in a Time of Great Change*, Oxford: Butterworth-Heinemann.

Earl, P.E. (1983). *The Economic Imagination: Towards a Behavioural Analysis of Choice*, Brighton: Wheatsheaf.

Earl, P.E. (1986). *Lifestyle Economics: Consumer Behaviour in a Turbulent World*, Brighton: Wheatsheaf.

Egan, J. (1998). *Rethinking Construction*, London: DETR.

Eichner, A. (1979). *A Guide to post-Keynesian Economics*, London: Macmillan.

Eichner, A. (1980). *The Megacorp and Oligopoly: Micro Foundations of Macro Dynamics*, New York: M.E. Sharp.

Felstead, A. (1993). *The Corporate Paradox: Power and Control in the Business Franchise*, London: Routledge.

Fine, B. (1975). 'Tendering strategy', in Turin, D. (ed.), *Aspects of the Economics of Construction*, London: Godwin.

Fine, B. and Leopold, E. (1993). *The World of Consumption*, London: Routledge.

Goold, M. and Campbell, A. (1987). *Strategies and Styles: The Role of the Centre in Managing Diversified Corporations*, Oxford: Blackwell.

Gordon, D.M. *et al.* (1994). 'Long swings and stages of capitalism', in Kotz, D.M. *et al.* (eds), *Social Structures of Accumulation*, Cambridge: CUP.

Gore, T. and Nicholson, D. (1991). 'Models of the land development process: a critical review', *Environment and Planning A*, 23, pp. 705–30.

Groak, S. (1992). *The Idea of Building*, London: Spon.

Gruneberg, S.L. (1989). 'Economic criteria', in Osbourn, D., *Components* (3rd edn), London: Mitchells.

Gruneberg, S.L. (1996). *Responding to Latham*, Ascot: CIOB.

Gruneberg, S.L. (1997). *Construction Economics: An Introduction*, London, Macmillan.

Gruneberg, S.L. and Weight, D. (1990). *Feasibility Studies in Construction*, London, Mitchells.

Harvey, A.C. and Jaeger, A. (1993). 'Detrending, stylised facts and the business cycle', *Journal of Applied Econometrics*, 8, pp. 231–41.

Harvey, D. (1982). *The Limits to Capital*, Oxford: Blackwell.

Harvey, D. (1989). *The Urban Experience*, Oxford: Blackwell.

Hay, D.A. and Morris, D.J. (1991). *Industrial Economics and Organisation: Theory and Evidence*, 2nd edn, Oxford: Oxford University Press.

Hayek, F. (1939). *Profits, Interest and Investment*, London: Routledge.

Heal, G. and Silberston, A. (1972). 'Alternative managerial objectives: an exploratory note', *Oxford Economic Papers*, 24, pp. 137–50.

Healey, P. (1992) 'An institutional model of the development process', *Journal of Property Research*, 9, pp. 33–44.

Healey, P. and Barrett, S. (1990). 'Structure and agency in land and property development processes: some ideas for research', *Urban Studies*, 27, pp. 89–104.

Hicks, J. (1973). *Capital and Time: A Neo-Austrian Theory*, Oxford: Clarendon.

Hillebrandt, P.M. (1984). *Analysis of the British Construction Industry*, London: Macmillan.

Hillebrandt, P.M. (1985). *Economic Theory and the Construction Industry*, London: Macmillan.

Hillebrandt, P.M. and Cannon, J. (eds) (1989). *The Management of Construction Firms*, London: Macmillan.

Hillebrandt, P.M. and Cannon, J. (1990). *The Modern Construction Firm*, London: Macmillan.

Hillebrandt, P.M., Cannon, J. and Lansley, P. (1995). *The Construction Company in and out of Recession*, London: Macmillan.

Hodgson, G. (1988). *Economics and Institutions: A Manifesto for a Modern Institutional Economics*, Cambridge: Polity Press.

Inter Company Comparisons (1982). *Business Ratios Business Ratio Report: Building and Civil Engineering (Major)*. London: Inter Company Comparisons.

Isard, W. (1942). 'A neglected cycle: the transport/building cycle', *Review of Economics and Statistics*, 24, pp. 149–58.

Ive, G. (1983). *Capacity and Response to Demand of the Housebuilding Industry*, London: UCL.

Ive, G. (1990). 'Structures and strategies: an approach to international comparison of industrial structures and corporate strategies in the construction industries', *Habitat International*, 14, pp. 45–58.

Ive, G. (1994). 'A theory of ownership types applied to the construction majors', *Construction Management and Economics*, 12, pp. 349–62.

Ive, G. (1995). 'Commercial architecture', in Borden, I. and Dunster, D. (eds), *Architecture and the Sites of History*, Oxford: Butterworth.

Ive, G. (1996). 'Innovation and the Latham Report', in Gruneberg, S. (ed.), *Responding to Latham*, Ascot: CIOB.

Ive, G. and Gruneberg, S. (2000) *The Economics of the Modern Construction Firm*, London: Macmillan.

Ive, G. and McGhie, W. (1983). 'The relation of construction to other industries and to the overall labour and accumulation process', *Production of the Built Environment, Proceedings of the Bartlett International Summer School*, London: UCL, 4, pp. 3-3–3-12.

Jackson, D. (1982). *Introduction to Economic Theory and Data*, London: Macmillan.

Janssen, J. (1983). 'The formal and real subsumption of labour to capital in the building process', *Production of the Built Environment, Proceedings of the Bartlett International Summer School*, London: UCL, 4, pp. 2-2–2.10.

Kalecki, M. (1971). *Selected Essays on the Dynamics of the Capitalist Economy*, Cambridge: Cambridge University Press.

Kay, N.M. (1984). *The Emergent Firm: Knowledge, Ignorance and Surprise in Economic Organisation*, London: Macmillan.

Keynes, J.M. (1936). *The General Theory of Employment, Interest and Money*, London: Macmillan.

Kindleberger, C. (1978). *Manias, Panics and Crashes: A History of Financial Crises*, New York: Basic Books.

Knight, F.H. (1933). *Risk, Uncertainty and Profit*, London: LSE.

Kotz, D.M., McDonough, T. and Reich, M. (eds) (1994). *Social Structures of Accumulation: The Political Economy of Growth and Crisis*, Cambridge: Cambridge University Press.

Kuznets, S. (1967). *Secular Movements in Production and Prices*, New York: A.M. Kelley.

Langlois, R.N. (1984). 'Internal organization in a dynamic context: some theoretical considerations', in Jussawalla, M. and Ebenfield, H. (eds), *Communication and Information Economics: New Perspectives*, Amsterdam: North-Holland.

Latham, M. (1994). *Constructing the Team*, London: HMSO.

Lee, F. (1999). *Post-Keynesian Price Theory*, Cambridge: Cambridge University Press.

Leopold, E. and Bishop, D. (1983). 'Design philosophy and practice in speculative housebuilding: parts 1 and 2', *Construction Management and Economics*, 1, pp. 119–44, 233–68.

Lewis, J.P. (1965). *Building Cycles and Britain's Growth*, London: Macmillan.

Linder, M. (1994). *Projecting Capitalism: A History of the Internationalisation of the Construction Industry*, London: Greenwood.

Lipietz, A. (1987). *Mirages and Miracles: The Crisis of Global Fordism*, London: Verso.

Littler, C. (1982). *The Development of the Labour Process in Capitalist Societies*, Aldershot: Gower.

Loasby, B.J. (1976). *Choice, Complexity and Ignorance: An Enquiry into Economic Theory and Practice of Decision Making*, Cambridge: Cambridge University Press.

Maddison, A. (1991). *Dynamic Forces in Capitalist Development*, Oxford: Oxford University Press.

Mandel, E. (1980). *Long Waves of Capitalist Development*, Cambridge: Cambridge University Press.

Marglin, S.A. (1974). 'What do bosses do? The origins and functions of hierarchy in capitalist production', *Review of Radical Political Economics*, 6, pp. 60–112.

Marglin, S.A. (1984). 'Knowledge and power', in Stephen, F.H. (ed.), *Firms, Organization and Labour*, London: Macmillan.

Marglin, S.A. and Schor, J.B. (1992). *The Golden Age of Capitalism*, Oxford: Clarendon Press.

Marx, K. (1970). *Capital: Volume 1*. (Trans. by Moore and Aveling from 3rd German edn; first published in English in 1887), London: Lawrence & Wishart.

Massey, D. and Catalano, A. (1978). *Capital and Land: Landownership by Capital in Great Britain*, London: Arnold.

Masterman, J.W.E. (1992). *An Introduction to Building Procurement Systems*, London: Spon.

Minsky, H. (1986). *Stabilising an Unstable Economy*, London: Yale University Press.

Morishima, M. (1984). *The Economics of Industrial Society*, Cambridge: Cambridge University Press.

Mueller, D.C. (1986a). *The Modern Corporation: Profits, Power, Growth and Performance*, Brighton: Wheatsheaf.

Mueller, D.C. (1986b). *Profits in the Long Run*, Cambridge: Cambridge University Press.

Mueller, D.C. (1990). *The Dynamics of Company Profits: An International Comparison*, Cambridge: Cambridge University Press.

National Economic Development Council (1976). *Construction into the Early 1980s: The Implication for Manpower and Materials of Possible Levels and Patterns of Demand*, London: HMSO.

National Economic Development Council (1978). *How flexible is construction?*, London: HMSO.

Neale, A. and Haslam, C. (1994). *Economics in a Business Context*, 2nd edn, London: Chapman & Hall.

Needham, D. (1978). *Economics of Industrial Structure, Conduct and Performance*; Eastbourne: Holt Rinehart & Winston.

Nelson, R. and Winter, S. (1982). *An Evolutionary Theory of Economic Change*, Cambridge, Mass.: Belknap.

North, D.C. (1990) *Institutions, Institutional Change and Economic Performance*, Cambridge: Cambridge University Press.

Office of National Statistics (Quarterly). *Economic Trends*, London: HMSO.

Okun, A.M. (1981). *Prices and Quantities – A Macroeconomic Analysis*, Oxford: Blackwell.

Park, W.R. (1979). *Construction Bidding for Profit*, New York: John Wiley.

Park, W.R. and Chapin, W.B. (1992) *Construction Bidding: Strategic Pricing for Profit*, Chichester: Wiley.

Penrose, E.T. (1980). *The Theory of the Growth of the Firm* (2nd edn), Oxford: Blackwell.

Punwani, A. (1997). 'A study of the growth – investment – financing nexus of the major UK construction groups', *Construction Management and Economics*, 15, pp. 349–61.

Rawlinson, S. and Raftery, J. (1997). 'Price stability and the business cycle: UK construction bidding patterns', *Construction Management and Economics*, 15, pp. 5–18.

Reich, M. (1994). 'How social structures of accumulation decline and are built', in Kotz, D.M. *et al.* (eds) *Social Structures of Accumulation*, Cambridge: CUP.

Reynolds, S. (1989). 'Kaleckian and post-Keynesian theories of pricing', in Arestis, R. and Kitromilides, Y. (eds), *Theory and Policy in Political Economy: Essays in Pricing, Distribution and Growth*, Aldershot: Edward Elgar.

Ricardo, D. (1973 edition). *On the Principles of Political Economy and Taxation*, London: Dent.

Richardson, G.B. (1960). *Information and Investment*, Oxford: Oxford University Press.

Robinson, J. (1933). *Economics of Imperfect Competition*, London: Macmillan.

Robinson, J. and Eatwell, J. (1973). *An Introduction to Modern Economics*, London: McGraw-Hill.

Rostow, W.W. (1971). *The Stages of Economic Growth*, Cambridge: Cambridge University Press.

Rowlinson, M. (1997). *Organisations and Institutions*, London: Macmillan.

Salter, W. (1969). *Productivity and Technical Change*, Cambridge: Cambridge University Press.

Sawyer, M. (1981). *Economics of Industries and Firms*, Beckenham: Croom Helm.

Schumpeter, J.A. (1934). *Theory of Economic Development*, Cambridge Mass.; Harvard University Press.

Shackle, G.L.S. (1952). *Expectations in Economics*, Cambridge: Cambridge University Press.

Shackle, G.L.S. (1961). *Decision, Order and Time in Human Affairs*, Cambridge: CUP.

Shackle, G.L.S. (1967). *Years of High Theory*, Cambridge: Cambridge University Press.

Shackle, G.L.S. (1968). *Expectations, investment and income*, Oxford: Clarendon.

Shackle, G.L.S. (1972). *Epistemics and Economics*, Cambridge: Cambridge University Press.

Sherman, H. (1991). *The Business Cycle*, Princeton: Princeton University Press.

Simon, H.A. (1955). 'A behavioural mode of rational choice', *Quarterly Journal of Economics*, 69, pp. 99–118.

Simon, H.A. (1957). *Models of Man*, New York: Wiley.

Simon, H.A. (1959). 'Theories of decision-making in economics and behavioural sciences', *American Economic Review*, 49, pp. 467–82.

Simon, H.A. (1976). *Administrative Behavior* (3rd edn), New York: Macmillan.

Simon, H.A. (1979). 'Rational decision making in business organizations', *American Economic Review*, 69, pp. 493–513.

Skinner, A.S. and Wilson, T. (1975). *Essays on Adam Smith*, Oxford: Oxford University Press.

Skitmore, M. (1989). *Contract Bidding in Construction*, Harlow: Longman.

Smith, A. (1976). *An Enquiry into the Nature and Causes of the Wealth of Nations*, Campbell, R.H. and Skinner, A.S. (eds), Oxford: Clarendon.

Smith, N. (1984). *Uneven Development*, Oxford: Blackwell.

Smith, T. (1992). *Accounting for Growth*, London: Century Business.

Smyth, H. (1985). *Property Companies and the Construction Industry in Britain*, Cambridge: Cambridge University Press.

Sraffa, P. (1926). 'The laws of returns under competitive conditions', *Economic Journal*, 36, pp. 535–50.

Stumpf, I. (1995). *Competitive Pressure on Medium Sized Regional Contractors and their Strategic Responses*, MSc Dissertation, London: Bartlett School, UCL.

Sugden, J.D. and Wells, O. (1977). *Forecasting Construction Output from Orders*, London: University College Environmental Research Group.

Sweezy, P. (1939). 'Demand under conditions of oligopoly', *Journal of Political Economy*, 47, pp. 568–73.

Thomas, B. (1972). *Migration and Urban Development: A Reappraisal of British and American Long Cycles*, London: Methuen.

Thompson, G. (1986). *Economic Calculation and Policy Formation*, London: Routledge.

Thurow, L.C. (1976). *Generating Inequality*, London: UCL.

Thurow, L.C. (1980). 'Education and economic equality', in King, J.E. (ed) *Reading in Labour Economics*, Oxford: OUP.

Turin, D. (1973). *Construction and development*, London: UCL.

Turin, D. (1969). *The Construction Industry: its Economic Significance and its Role in Development*, UNIDO, London.

Turin, D. (ed.) (1975). *Aspects of the Economics of Construction*, London: Godwin.

Tylecote, A. (1992). *The Long Wave in the World Economy*, London: Routledge.

Van Duijn, J.J. (1983). *The Long Wave in Economic Life*, London: Allen & Unwin.

Varoufakis, Y. (1998). *Foundations of Economics*, London: Routledge.

Veblen, T. (1921). *The Engineers and the Price System*, New York: Harcourt Brace.

Veblen, T. (1964). *The Instinct of Workmanship*, New York: Augustus Kelly.

Vickrey, W. (1961). 'Counterspeculation, auctions, and competitive sealed tenders', *Journal of Finance*, 16, pp. 8–37.

Williamson, J. (1966). 'Profit, growth and sales maximisation', *Economica*, 33, pp. 253–6.

Weisskopf, T.E. (1994). 'Alternative social structure of accumulation approaches to the analysis of capitalist booms and crises', in Kotz, D.M. *et al.* (eds) *Social Structures of Accumulation*, Cambridge: CUP.

Williamson, O. (1975). *Markets and Hierarchies*, New York: Free Press.

Williamson, O. (1985). *The Economic Institutions of Capitalism*, New York: Free Press.

Wilson, T. and Andrews, P.W.S. (eds) (1951). *Oxford Studies in the Price Mechanism*, Oxford: Clarendon Press.

Winch, G.M. (1986). 'The labour process and labour markets in construction', *International Journal of Sociology and Social Policy*, 6, pp. 103–16.

Winch, G.M. (1989). 'The construction firm and the construction project: a transaction cost approach', *Construction Management and Economics*, 7, pp. 331–45.

Winch, G.M. (1995). 'Project management in construction: towards a transaction cost approach', *Le Groupe Bagnolet Working Paper*, 1, London: Bartlett School of Graduate Studies, UCL.

Winch, G.M. (1996a). 'The contracting system in British construction: the rigidities of flexibility', *Le Groupe Bagnolet Working Paper*, 6, London: Bartlett School of Graduate Studies, UCL.

Winch, G.M. (1996b). 'Contracting systems in the European construction industry', in Whitley, R. and Kristensen, P. (eds), *The Changing European Firm: Limits to Convergence*, London: Routledge.

Winch, G.M. and Schneider, E. (1993). 'Managing the knowledge-based organisation: the case of architectural practice', *Journal of Management Studies*, 30, pp. 923–37.

Winch, G.M. and Campagnac, E. (1995). 'The organisation of building projects: an Anglo-French comparison', *Construction Management and Economics*, 13, pp. 3–14.

Winch, G.M. (1998). 'The growth of self-employment in British construction', *Construction Management and Economics*, 16, pp. 531–42.

Wolfson, M.H. (1994). 'The financial system and the structure of accumulation', in Kotz, D.M. *et al.* (eds) *Social Structures of Accumulation*, Cambridge: CUP.

Wood, A. (1975). *Theory of Profit*, Cambridge: Cambridge University Press.

Index